Contributions to Finance and Accounting

The book series 'Contributions to Finance and Accounting' features the latest research from research areas like financial management, investment, capital markets, financial institutions, FinTech and financial innovation, accounting methods and standards, reporting, and corporate governance, among others. Books published in this series are primarily monographs and edited volumes that present new research results, both theoretical and empirical, on a clearly defined topic. All books are published in print and digital formats and disseminated globally. This book series is indexed in Scopus.

Musa Gün · Burcu Kartal
Editors

Machine Learning in Finance

Trends, Developments and Business Practices in the Financial Sector

Springer

Editors
Musa Gün
The Faculty of Economics
and Administrative Sciences
Recep Tayyip Erdoğan University
Rize, Türkiye

Burcu Kartal
The Faculty of Economics
and Administrative Sciences
Recep Tayyip Erdoğan University
Rize, Türkiye

ISSN 2730-6038 ISSN 2730-6046 (electronic)
Contributions to Finance and Accounting
ISBN 978-3-031-83265-9 ISBN 978-3-031-83266-6 (eBook)
https://doi.org/10.1007/978-3-031-83266-6

This Springer imprint is published by the registered company Springer Nature Switzerland AG
The registered company address is: Gewerbestrasse 11, 6330 Cham, Switzerland

Preface

The integration of machine learning into the dynamic and rapidly evolving field of finance has ushered in a new era. In this book, we aim to explore the impact of machine learning on financial applications and provide insights into how advanced techniques are transforming the industry.

Machine learning has attracted the financial industry's attention due to its ability to analyze large amounts of data and apply different analyses than traditional methods. Machine learning applications, such as financial transformation, portfolio optimization, digital currencies, and exchange rate forecasting, are vast and constantly expanding.

This book is designed for finance professionals, data scientists, students, and those who want to understand the implications of machine learning algorithms in finance. We provide theoretical foundations and practical insights, making complex concepts accessible to readers with different levels of expertise.

We are extremely grateful to the many individuals and organizations that supported our efforts. We would like to thank our family and friends whose encouragement and feedback helped to shape this work. And, we are especially grateful to our colleagues for sharing their invaluable expertise in finance and machine learning algorithms.

The book is enriched by the contributions of researchers who have integrated machine learning into finance. Their pioneering efforts and innovative ideas have created a rich body of knowledge.

We would also like to thank the first readers and reviewers for their constructive criticisms. Their different perspectives have enriched the content of the book.

Please take a critical and enquiring approach when analyzing the papers. The field of machine learning in finance is still in its infancy, and there is much to explore and discuss. We believe that when those working in this field interact, this field will develop.

It is thought that the book will make a serious contribution to the literature. We dedicate the book to anyone who wants to harness the potential of machine learning in finance. And we wish it to be a valuable resource and source of inspiration.

Rize, Türkiye

Musa Gün
Burcu Kartal

Acknowledgments The authors would like to thank their families for their help and patience during the writing of this book. Their support was instrumental in making this book a reality.

The authors also acknowledge the valuable contribution of the reviewers in improving the quality, consistency, and content presentation of the chapters.

About This Book

The aim of this book is to identify important issues related to the use of machine learning algorithms in finance.

For this purpose, all different perspectives of finance, such as financial transformation, portfolio optimization, digital currencies, and exchange rate forecasting, are analyzed with machine learning algorithms. By focusing on different perspectives, it may be possible to produce more appropriate strategies for the development of finance with machine learning algorithms.

Finance makes a significant contribution to the development of economies. Machine learning algorithms will also contribute to the development of finance and improve economies. The target audience and potential users of this book are defined below.

- Researchers
- Academics
- Policymakers
- Government officials
- Senior executives of companies

Contents

About the Editors

Musa Gün is Associate Professor of Finance at Recep Tayyip Erdoğan University. Before this position, he worked in different financial and audit positions at Deloitte Touche Tohmatsu (İstanbul office), at Finansbank (outsourcing and project management) in Istanbul, and at BIM Stores (internal audit) in Istanbul—Türkiye. Dr. Gün has a BS in International Trade (in English) from Boğaziçi University—which has a history of over 150 years and is widely recognized as one of the world's most prestigious universities (2005). He got his master's degree in Financial Economics from Dogus University (2010). He also has a Ph.D. in Finance from Istanbul University (2015). His research interests lie in banking, finance, and financial crises. His research also has significantly advanced the understanding of asset pricing models, market anomalies, credit risk management, investment valuations, and financial technologies (Business Administration Department, Recep Tayyip Erdoğan University, musa.gun@erdogan.edu.tr).

Burcu Kartal is Assistant Professor in the Department of Quantitative Methods at Recep Tayyip Erdoğan University. She completed her undergraduate education at Karadeniz Technical University with a double major in Statistics and Computer Science and Business Administration (2006). She obtained her master's degree in Business Administration and completed her Ph.D. at Istanbul University in the Department of Quantitative Methods (2015). Her areas of interest include quantitative research methods, machine learning, artificial intelligence, heuristic algorithms, and deep learning. She has been working for 15 years in the Department of Quantitative Methods (Business Administration Department, Recep Tayyip Erdoğan University, burcu.kartal@erdogan.edu.tr).

Machine Learning in Finance: Transformation of Financial Markets

Musa Gün

Abstract This study explores the transformative role of machine learning in the financial sector, highlighting its evolution, methodologies, and diverse applications. Driven by the force of improvements in artificial intelligence, machine learning has fundamentally transformed traditional finance such that systems can now utilize big data, identify patterns, and make data-driven decisions with little to no human input. The study describes core types of machine learning: supervised learning, unsupervised learning, semi-supervised learning, and reinforcement learning with application examples in algorithmic trading, credit scoring, fraud detection, customer service, portfolio management, and insurance underwriting. Algorithmic trading, predictive analytics, and robo advisors turned possible by machine learning have made capital markets more efficient while delivering personalized financial services and security mechanisms with the help of machine learning. However, adopting machine learning raises ethical, regulatory, and data privacy issues. Focusing on the challenges and advancements, this paper offers a detailed examination of how machine learning reshapes financial markets while laying the groundwork for future work to make sense of its growing complexity.

1 Introduction

Mitchell (1997), a pioneer in machine learning, defines machine learning as the ability of machines to acquire experience and improve machines' performance based on similar experiences. Machine learning, a form of artificial intelligence, allows hardware systems to process data, search out patterns, and forecast future outcomes (Mitchell 2006). In this context, machine learning refers to systems of algorithms that learn from data sets and perform better as they acquire new data over time.

Learning refers to the process of acquiring knowledge and improving performance through experience. This discipline is a relatively new field that is rapidly growing

M. Gün (✉)
The Faculty of Economics and Administrative Sciences, Recep Tayyip Erdogan University, Merkez, Rize, Türkiye
e-mail: musa.gun@erdogan.edu.tr

M. Gün and B. Kartal (eds.), *Machine Learning in Finance*, Contributions to Finance and Accounting, https://doi.org/10.1007/978-3-031-83266-6_1

and developing. It especially focuses on developing software techniques to transform data into actionable insights. Essentially, machine learning functions as a data exploration method that automates the creation of analytical techniques and is a branch of artificial intelligence where systems can recognize patterns and make judgments without human interaction thanks to learning from existing data.

A machine learning approach considers various design considerations, including the type of training data, the goal function to be learned, how this function is represented, and the specific learning algorithm used. A machine learning task requires a well-defined task, an effective evaluation metric, and appropriate data resources. This field focuses on two essential questions: "How can we build computer systems that can improve automatically?" and "What are the fundamental theoretical principles that govern all learning systems?" Machine learning is significant for its practical applications and contribution to scientific inquiry. As a subfield of computer science aimed at programming machines to learn, machine learning can also be viewed as a branch of artificial intelligence due to its capacity to identify patterns akin to human intelligence. Unlike traditional artificial intelligence, which often seeks to replicate intelligent behavior, machine learning aims to perform tasks that exceed human capabilities by leveraging the strengths of computers (Shalev-Shwartz and Ben-David 2014).

Machine learning utilizes mathematical disciplines such as probability, statistics, and optimization theory to uncover patterns in data. Within this framework, the machine learns from historical data to build a model that generates future predictions. Unlike traditional software engineering, where developers specify patterns, machine learning algorithms autonomously discover these patterns (Grigorev 2021). Machine learning reveals trends and hidden relationships by harnessing statistical methods and powerful computing capabilities, facilitating accurate predictions.

In this direction, the chapter investigates the idea of machine learning, offers a short history, shows how it is used in finance, discusses some changes that machine learning has generated in the financial system, and provides an overview for future research on this.

The rest of this chapter is organized as follows: Section Two provides a short history of machine learning; Section Three presents the different types of machine learning approaches; Section Four outlines the examples of machine learning applications in finance; and the last Section concludes the study.

2 A Short History of Machine Learning

The influential work "Computing Machinery and Intelligence" written by Alan Turing in 1950 forms the basis of the notion of machine learning (Turing 2009). That study explores the question "Can machines think?" by examining interactions between a human, a judge, and a machine, ultimately proposing that machines could demonstrate human-like thinking. Nine years later, Arthur Samuel from IBM company promoted the idea of machine learning in his paper. The study aims to introduce a self-learning program for checkers (Samuel 1959). In the same year, Cahit Arf emphasized the capacity of machines to perform logical and analytical tasks (Arf 1959). A significant milestone in machine learning occurred in 1962 when expert checkers player Robert Nealey lost a match to an IBM computer, marking a pivotal moment in the development of the field.

The latter part of the 1900s is generally called a transformative "golden age" for machine learning, marked by major advancements that have reshaped the ground. In 1957, Frank Rosenblatt introduced the first neural network to simulate human cognitive processes. A decade later, in 1967, Cover and Hart developed the "Nearest Neighbor" algorithm, a fundamental technique for solving data classification tasks (Cover and Hart 1967). Gerald DeJong introduced the explanation-based learning system in 1981, which enables computers to examine and interpret training inputs, disregard irrelevant information, as well as create generalized rules.

In 1985, a significant milestone in machine learning emerged with the founding of AT&T's research group. By 1992, Ramesh and Wilpon developed the first automatic speech recognition system, utilizing a novel approach known as Hidden Markov Models (Ramesh and Wilpon 1992). In 1995, Cortes and Vapnik introduced the support vector machine technique, revolutionizing large-scale data classification (Cortes and Vapnik 1995). That same year, LeCun, Bottou, Bengio, and Haffner introduced a convolutional neural model, pioneering in enhancing recognition procedures (LeCun et al. 1998). In 1996, Freund and Schapire's AdaBoost algorithm advanced unstructured data processing through decision tree methods (Schapire 2013).

By the first half of the twenty-first century, machine learning has emerged as a critical field, driving numerous advancements. In 2001, natural language processing was integrated with machine learning to create interactive voice response systems, enhancing human–computer interactions. By 2011, researchers in deep neural networks developed algorithms capable of training models on millions of examples, marking a significant leap in model scalability. Google and Facebook made machine learning a core component of their systems in early 2014. By 2015, AT&T's speech and language technology had become the leading transformative approach to customer services. Interactions grew even bigger in 2017 with the emergence of Digital Roots, a company that uses machine learning-based services to help marketers quickly communicate with their social media followers (Janiesch et al. 2021).

With the launch of the principles of artificial intelligence (AI), the Organization for Economic Cooperation and Development (OECD) created the first worldwide framework for the regulation of reliable artificial intelligence in 2019 (OECD 2019). The OECD report for 2021–2022 analyzes the integration of machine learning and artificial intelligence. It focuses on how these new technologies transform the financial industry and business models. The report emphasizes the widespread use of machine learning in finance, highlighting and detailing its significance in managing risks, credit operations, algorithms for trading, solutions based on blockchain technologies, mobile financial services, chatbots, credit scoring, fraud detection, and insurance practices. In essence, the OECD's principles offer a policy framework that prioritizes equitable and sustainable growth, and social welfare, guiding both governments

Table 1 OECD's key areas for AI principles

Areas	Explanation
Equitable and sustainable growth, and social welfare	AI should serve the well-being of both society and the environment by fostering sustainable development and inclusive economic growth
Humanistic values and equity	AI systems should protect human rights, advocate justice, and encourage democratic rules with accountability and transparency
Transparency and explainability	AI systems should be intelligible to all key stakeholders and feature clear and transparent methods for making decisions and producing outputs
Robustness, security and safety	AI systems should be trustworthy and secure in various applications, which involves mechanisms to prevent threats and unexpected consequences
Accountability	To ensure accountability at every stage, those in charge of developing and implementing AI systems should abide by ethical and legal norms

and private-sector entities to manage artificial intelligence technologies properly. The principles outline five major themes summarized in Table 1.

The above notions have shaped international regulatory debates, such as the European Union's planned act on AI, and promoted global consensus on moral principles and governance best practices. To guide and contribute to the development of appropriate international AI standards, the OECD continues to evaluate the effects of AI through regular publications that offer insights into AI applications in industries including public services, healthcare, and finance.

3 Machine Learning Categories

To reiterate, machine learning refers to the processes by which a computer program may improve its performance by becoming familiar with certain duties and success criteria (Jordan and Mitchell 2015). The creation of analytical models that are capable of cognitive tasks like recognizing objects, comprehending, and translating natural language translation automatically is the main objective of machine learning systems. By using algorithms that systematically acquire problem task-specific data for training, machines can uncover complicated patterns of hidden knowledge without the need for explicit programming (Bishop 2006).

Machine learning is especially effective in solving problems that involve high-dimensional data contents like classification, clustering, and regression. Through iterative learning from past computations and detecting patterns within extensive datasets, machine learning enables the generation of reliable, consistent decisions.

Consequently, machine learning algorithms are widely applied across diverse fields of finance, namely credit scoring, fraud detection, best offer analysis, and recognition of audio and pictures.

The main focus of machine learning is enabling computer systems to acquire knowledge from information data and improve over time. The literature generally categorizes machine learning approaches into four distinct groups: supervised, unsupervised, semi-supervised, and reinforcement learning. The approaches and application areas of each category are further explained.

3.1 Supervised Learning Approach

The supervised learning approach focuses on predicting correlations between inputs and outputs in labeled datasets. This approach uses datasets that connect input data with corresponding target outputs to discover underlying patterns. Each target output is linked to the inputs by executing a mapping function. The labeled datasets can be used and analyzed by supervised learning and predictions can be generated for the current and comparable samples. The two main duties in this technique are classification and regression. This learning process continues whenever the model gets the desired degree of accuracy. This facilitates the system to properly map the inputs and outputs.

The classification algorithms assign data to predefined categories in this learning approach. On the other hand, the main goal of computational methods in regression analysis is to try to predict dependent variables. Methodologies such as logistic regression, random forest learning, and decision tree modeling are the major methods of supervised learning approaches (Mohri et al. 2018).

3.2 Unsupervised Learning Approach

The primary weakness of the supervised learning approach is that it requires fully labeled data. Therefore, the supervised learning approach cannot successfully handle unclassified or unlabeled observations. The unsupervised learning approach developed in this context is often applied in cases where the data is unlabeled. Data mining functions, especially those that contain large amounts of unstructured data, can be given as examples.

The unsupervised learning method is used to find out related patterns from a data set. It does not need direct feedback but evaluates the probability of certain patterns appearing in your data and creates a model accordingly. Since unsupervised learning is not executed according to any known logical procedural steps, it will be a more complex process than supervised learning (Jo 2021; Watt et al. 2020) Besides, unsupervised learning shows that through mathematical ways of representing patterns, similarities, and dissimilarities rather than using correct and incorrect outcomes. In short, unsupervised learning algorithms often find application in clustering, association analysis, and anomaly detection.

3.3 Semi-supervised Learning Approach

The semi-supervised learning method combines a large set of unlabeled data with a small quantity of labeled data. It is used for classification and regression analysis.

This approach is useful due to the easy availability of unlabelled data while it is far costlier or harder to get labeled data. Unlike supervised learning, which predicts a few unknown data points with tons of labeled data, the semi-supervised approach is the opposite. To repeat, the semi-supervised algorithm is one in which a small number of examples are annotated along with a large collection of unlabeled instances for prediction. When viewed through this framework, it maintains the fence between supervised and unsupervised learning to leverage available data better (Srinivas et al. 2021).

3.4 Reinforcement Learning Approach

Reinforcement learning is a branch of machine learning utilized for decision-making in dynamic environments. These dynamic environments are typically driven by trade and investment management issues. In contrast to the passive learning of labeled data in supervised learning, decision-making in reinforcement learning is made under uncertainty and with incomplete feedback. Once this system is trained, it learns from interacting with the environment; it does not need separate training data and can use any environment.

This approach makes it particularly useful for problems such as portfolio optimization, derivative pricing, and market-making. Moreover, reinforcement learning is appropriate once models diverge from real-world dynamics. This is an approach centered around doing the best possible job to maximize cumulative reward and minimize risk within particular operations of machines (Mohammed et al. 2017). Reinforcement learning, inspired mainly by concepts in supervised learning and dynamic programming, evaluates an action as successful or failed based on the feedback gained if the goal is not met. It also stops errors from being repeated and leads to an increase in future workflow results. The problem of reinforcement learning is a hundred times more difficult and complex than supervised or unsupervised learning (Sarker 2021).

4 Applications of Machine Learning in Finance

With the growing amount of data available and advancements in computing power, studies, and new discussions are emerging today that confirm machine learning is relevant to finance. So, machine learning is also a very important concept in finance. Its benefits in modern finance are undeniably one of the most productive technologies. It can process vast volumes of information in a short time, and automate to a great extent the decision-making processes.

The machine learning approach is a useful solution in many applications that include trading algorithms, predictive modeling, and risk management. Additionally, it is widely utilized for fraud detection, portfolio management, and insurance determination.

4.1 Algorithmic Trading

Algorithmic trading is the subfield of automated, black-box, or robo-trading that deals with developing and deploying computer algorithms to execute trades across one or more financial markets (Kissell 2014). It includes different forms of electronic trading, including smart routing, program trading, and rule-based systems, based on the type of algorithm used (Domowitz and Yegerman 2006). Hendershott et al. (2011) define it as using algorithms to manage efficiently and route orders, providing speed-centric market access.

This type of trading uses automated financial decision-making algorithms to make trading and investment decisions, price assessments, and risk assessments over a range of instruments such as stocks, bonds, and derivatives. Investors express their objectives as "numbers" and financial managers as mathematicians to build algorithms to select assets, markets, and portfolios. The algorithms make the price, time to sale, and number of shares as optimal as possible with a completely autonomous system.

The execution may vary depending on the way a system is designed. Some are fully automated and can control all steps in a trading process, while others help human traders and focus on automating decision support. Best suited to systems with more readily understandable mathematical models where having a correct parameter (start times, order types, and duration) definition is the key point for success. Such systems interact with trading platforms, process market data in real-time, and send out buy or sell instructions based on trading algorithms (Chaboud et al. 2014).

The literature highlights several notable advantages of algorithmic trading. Table 2 provides a brief overview.

Research studies on algorithmic trading emphasize its impact on market volatility, trading efficiency, and the potential risks associated with financial markets. As technology advances, factors such as connection speeds and colocation (the physical proximity of trading servers to exchanges) have been recognized as essential elements

Table 2 Key benefits of algorithmic trading

Advantages	Brief overview
Reduced commissions	Transaction costs are lower than traditional intermediaries. It allows investors to pay lower commissions by transmitting their orders directly to exchanges without intermediaries
Privacy and anonymity	It means that transactions are carried out anonymously and confidentially. This prevents information leakage
Improved velocity and accessibility	It means being able to access multiple markets simultaneously at high speed, which makes transactions more transparent.

that enhance the speed and efficiency of algorithmic trading. In their study, Domowitz and Yegerman (2006) concluded that algorithmic trading lowers transaction costs compared to traditional methods and that performance efficiency grows with growing transactions.

Gsell (2008) studied the impact of algorithmic trading on volatility and concluded that high-volume trading contributes to volatility, while low latency has a positive effect on market stability. Additionally, instances like the Flash Crash (a type of stock market crash) demonstrated that algorithms can enhance shocks to the market, causing abrupt price changes. Prix et al. (2007) cited that algorithmic trading also brings new unobserved patterns into market structures.

4.2 *Credit Risk Assessment and Scoring*

In this credit analysis process, one cannot underestimate the importance of prediction and classification tasks as these depend on various machine learning algorithms. These algorithms calculate the likelihood of an approval or rejection of a loan request by building models for predictive analysis based on historical data patterns. Below are a few of the most requested algorithmic solutions within this area (Popovych 2022):

K-NN (K Nearest Neighbor) algorithm identifies the characteristics of data based on their nearest neighbors in the training set and thus classifies them. It was developed in the 1950s but took off when computational power became more powerful. The K-NN algorithm is a credit estimation model that compares the profiles of their applicant with profiles such as similar cases from earlier (Han and Kamber 2006).

Naive Bayes algorithm is a classification technique based on Bayes' Theorem with an assumption of independence among predictors. It can estimate based on previously proven facts in the dataset for credit risk analysis to assume features as independent of each other and equally important to predict/deal with non-payment by a customer due to their credit history. This makes it particularly well-suited for classifying loan applications, and predicting how likely a certain result is based on the class probabilities (Boschetti and Massaron 2018).

Decision Tree algorithm is an important algorithm in machine learning and data mining. It splits the data based on certain criteria and branches the dataset. It can be used for classification or regression problems. Nodes and branches are the fundamental aspects of a decision tree; every node is a decision point (or classification), while each branch separates the data into subsets based on these decisions (Williams 2011).

Decision trees are widely used for example in credit risk analysis, classification, and regression problems. It is a perfect approach in which loan applications are assessed based on historical credit risk data.

LDA, or linear discriminant analysis, is an algorithm for classifying data by using a linear boundary to separate classes. Named after Ronald A. Fisher, this algorithm uses linear combinations of features to optimize class discrimination. For credit analysis, for example, LDA can discriminate loan repayment outcomes into two classes of data: loans that will be paid back and loans that will not. LDA proves more useful for binary classification problems as it lets to achieve higher accuracy.

Logistic Regression algorithm is mainly used for binary classification and it is useful to estimate the probability of whether a loan request will get approved or rejected. Credit risk assessment examines the relationship between an applicant's fiscal traits and the possibility of loan defaulting. Due to its functionality to analyze the effects of independent variables on loan affordability, logistic regression has been a commonly accepted method in credit risk analysis.

The Artificial Neural Network algorithm, which takes inspiration from how the human brain functions, is effective in learning and classifying large, complex datasets for credit analysis purposes. The power of neural is the finding of these invisible patterns existing within large data, which makes it well-suited for long-term forecasting models.

ZeroR algorithm, the most rudimentary classification method simply predicts the majority class and severs a baseline for performance comparison. ZeroR is frequently used as a baseline for credit analysis due to its simplicity, serving as an effective reference point against which to measure the accuracy of more complex algorithms.

These types of algorithms are used to evaluate elements affecting loan affordability, risk assessment, and accurate predictions. Furthermore, the performance of every algorithm depends on data and task, therefore it is imperative to choose an appropriate algorithm for the proper data and classification objectives.

4.3 *Fraud Detection and Prevention*

In banking, fraud, and forgery pose not only one of the most serious security risks but also result in losses that amount to billions each year, legal matters due to lack of control, and revocation with customers. To protect the integrity of institutions, fraud detection, and prevention are important. There are various types of fraud such as phishing, credit card skimming, telephone marketing scams, identity theft, fake cheques, mobile banking frauds, ransomware, and false financial reporting (Salvin and Lepcha 2022). There are other types as well, which include asset misappropriation (theft or misuse of records) and corruption (bribery and extortion). When related to cash assets, schemes such as larceny, skimming, or false invoices/ reimbursements are referred to as cash asset misappropriation. Such criminal acts are actual or attempted fraud perpetrated by customers, vendors, or third parties.

One of the most popular security measures is a fraud detection system that is usually discovered in financial institutions and card payment networks. Its main utility is to detect unauthorized transactions, stolen or duplicated credit cards, and other frauds. Fraud detection systems are important for combating fraudulent actions with increasing online shopping and the trend of digital payment methods (Mishra and Pandey 2021).

Artificial intelligence and data analytics algorithms are among the key strategies in fraud detection. Commonly used techniques include the neural network, support vector machine, logistic regression, fuzzy logic system, local outlier factor, isolation forest, and k-nearest neighbor (Yang et al. 2009). They use automatic analysis of patterns in data to detect anomalies connected with suspicious activity and identify new potentially fraudulent transactions.

Abnormal user behaviors are detected through fraud detection systems that learn typical patterns within the population, which enables them to flag any suspicious activity, such as large transfers, geographical inconsistency, or unauthorized activities (Bin Sulaiman et al. 2022). They also combat identity theft, fake identities, and phishing by automating fraud prevention (like stopping suspicious transactions). However, they can generate false positives which necessitates frequent updates and also manual inspection. Modern systems intend to limit these errors with higher security and a better customer experience.

4.4 Customer Service

From phone calls and emails to text messages, machine learning is revolutionizing customer service with extensive data being analyzed. It is a technology that powers 24/7 automated chatbots, and virtual assistants which can resolve routine inquiries without human intervention, allowing human agents to focus on more complex issues, thus improving service quality and customer experience (Sabbeh 2018; Borg et al. 2021; Wu et al. 2022). Such tools use natural language processing that quickly responds to specific customer requests. Using data from the past, predictive analytics is more often used to predict customer behavior (like churn) or to recommend personalized experiences.

To unlock the optimum potential of machine learning in customer care, organizations must track and tune model performance, maintain data quality, and embed models into pre-existing support systems. Data privacy and security come first, and there should be clear objectives and performance metrics to ascertain success. In the end, machine learning converts customer experience into a competitive business by enabling quicker and more personalized interactions between organizations and customers.

4.5 Portfolio Management and Robo-Advisors

Asset management firms use robo-advisors to help create investing strategies that suit the customer's financial goals and risk tolerance. These technologies are being increasingly adopted to provide more specific and efficient solutions to customers. Portfolios created by robo-advisors, focus on age, income level, or the resources you have already accumulated and your financial goals. They then direct users to the asset classes and financial instruments closest to their preferences and goals. Additionally, these portfolios are designed to adapt quickly to shifts in market conditions or changes in users' goals. These capabilities help robo-advisors optimize their investments and manage their portfolios for the user autonomously, without a human advisor.

The literature contains many studies about the successful application of machine learning techniques for financial forecasting and portfolio selection. Most of these are stock price prediction, market trend forecasting, and portfolio optimization. The studies evaluate the accuracy of these techniques and compare the performance of different algorithms.

Commonly used methods in studies include artificial neural networks, support vector machines, decision trees, and LSTM (Long Short-Term Memory). Indeed, these methods are frequently employed to predict index trends or price variations of individual shares (Fischer and Krauss 2018; Wang et al. 2020). This attempts to compare the accuracy rates of individual machine-learning methods to explore which model provides better performance (Shen et al. 2018).

In other words, the studies regarding machine learning on portfolio selection focus on the construction of portfolios that outperform market portfolios in terms of return while being validated in terms of risk-return ratios (Paiva et al. 2019). The hybrid model approach combines the merits of machine learning algorithms to achieve an improved forecasting performance. In summary, the literature shows that using machine learning in finance has great promise as it provides investors with sophisticated tools that potentially lead to more informed decision-making (Ta et al. 2020). The existing research showed that machine learning helps to increase the accuracy of forecasts, making better predictions in portfolio selection (Boudabsa and Filipović 2022).

4.6 Sentiment Analysis

Sentiment analysis is a subfield research discipline of computer science focused on identifying and categorizing emotional attitudes toward a text. It refers to the automatic identification and classification of sentiment expressions in text from different sources like social media, consumer as well as employee feedback, surveys, and product reviews (Cambria et al. 2017). Sentiment analysis helps financial analysts and investors to buy shares or avoid them by analyzing news articles or posts on social media platforms about specific stocks or firms in the field of finance.

Sentiment analysis utilizes numerous approaches in detecting and classifying feelings embedded in texts which are mainly grouped into two major categories; machine-learning methods, and human-directed dictionary-based methods (Farhadloo and Rolland 2016; Agarwal et al. 2020).

Machine Learning Approaches:

- *Training Data*: These methods create models from previously labeled datasets, which they then apply to sample texts according to the emotional context.
- *Algorithms*: The algorithms cover support vector machines, decision trees, naive Bayes, neural networks, and more traditional deep learning algorithms.
- *Deep Learning*: Methods such as recurrent neural networks and long short-term memory models allow more complex feature extraction and support better comprehension of co-occurrence in texts.

Dictionary-Based Approaches:

- *Emotion Dictionary*: Using a predefined emotion dictionary where words are categorized into emotional classifications such as positive or negative and neutral to infer the sentiment classification of the documents.
- *Scoring*: Emotional scores are calculated by matching the content word in a text with that corresponding to them in the dictionary. These totaled scores are used to calculate the text's overall sentiment.
- *Advantages and Limitations*: The dictionary-based approaches are fast and easy to implement. However, they commonly have issues with analyzing the sub-textual nature of language (like irony or metaphor) because they cannot keep in mind the contextual meaning of words.

Thus, sentiment analysis approaches are extensively used to capture emotion-related content in the text data. There are many appropriate ways to statistically analyze distributions, and the choice of which method is applied will vary depending on both the analysis goals and data properties. An appropriate assessment and implementation of these methods is, therefore, necessary for the effective classification of sentimental information (Cambria et al. 2017).

4.7 Insurance Underwriting

As one of the most data-focused industries, insurance has made use of the benefits that artificial intelligence can bring to its business model over recent years of rapid digitalization. Companies in the insurance sector are particularly experiencing changes with artificial intelligence and machine learning technologies impacting key areas of their business, such as sales, customer service, claims management, underwriting, and pricing, smartly creating extensive operational efficiencies.

Some leading insurance companies have begun utilizing AI with advanced technologies and are investing heavily in these digital solutions. Such technologies include predictive analytics, deep learning, the Internet of things, and natural language processing; even virtual assistants and robotic process automation are used for trade (Venkatasubbu et al. 2023). For instance, AXA Insurance uses Google TensorFlow to predict traffic accidents, and Lemonade leverages virtual assistants for fraud detection and customer engagement.

As per the Value Chain Model of Michael Porter, the use of AI technologies improves the processes infrastructures, risk, and premium evaluations enabling insurance companies to perform efficiently (Eling and Lehmann 2018). Both core activities (sales and underwriting) and supporting activities (IT, legal, etc.) are optimized with these technologies in the insurance value chain. Digitalization especially leads to the expansion of online distribution channels, the analysis of customer data, and the processing of information collected from mobile devices. It helps companies analyze customer behavior, detect fraud, and do risk scoring with the help of AI technologies (Sahu 2019). Top insurers including AXA, Lemonade, Fukoku, Lumnion, Aetna, and AIG are using AI and machine learning technologies such as NLP, visual analysis, predictive analytics deep learning, and Internet of things to provide customer-centric experiences and automated services. In short, AI and machine learning provide a wide range of benefits to the insurance industry enabling operational efficiency, cost reduction, fraud detection and prevention, and enhanced customer experience among others. i.e., besides the fact that AI and machine learning are changing old business models, making them more competitive, challenged with a new paradigm of:—"best technology = best assurance"(Corea 2019).

5 Conclusion

This paper takes a wider, holistic view and certainly helps understand how machine learning changes finance. Its framework studies fundamental sector methods and applications and their impacts on the sector and highlights academic work with a vision toward the future of finance.

Machine learning, a branch of artificial intelligence that involves systems using algorithms to analyze large datasets, identify trends, and make predictions with little human involvement, is changing the face of finance. This study describes the principal types of machine learning—supervised, unsupervised, semi-supervised, and reinforcement learning—and their specific applications in finance. In algorithmic trading, markets driven by machine learning models and artificial intelligence can make trades in milliseconds, ensuring higher speed and efficiency than human-driven trades. Machine learning also plays a critical role in risk assessment, credit scoring, and fraud detection by enabling financial institutions to make data-driven decisions that improve security while reducing costs and improving customer experience.

By redefining how financial operations are carried out, machine learning technologies and software solutions push the financial industry toward inclusiveness and accessibility while increasing efficiency. Also, machine learning, data-driven insights, and automation capabilities can help financial institutions across the decision-making continuum, focusing on improving operational efficiency and customer personalization. But these progressives are not without challenges: machine learning's dependence on massive datasets raises data privacy and security issues.

To wrap it up, the paper elucidates that machine learning will drive groundbreaking change in contemporary finance, moving us toward a future of increasing financial autonomy and allowing for more data-driven and user-oriented financial ecosystems. Despite regulatory, ethical, and technical challenges that must be addressed, machine learning promises a future of finance that is more open, transparent, and bespoke per person, restructuring industry fundamentals while raising the bar for global financial innovation.

References

Agarwal B, Nayak R, Mittal N, Patnaik S (eds) (2020) Deep learning-based approaches for sentiment analysis, vol 12. Springer, Singapore, p 319

Arf C (1959) Makine Düşünebilir mi ve Nasıl Düşünebilir? Atatürk Üniversitesi 1958–1959 Öğretim Yılı Halk Konferansları 1:91–103

Bin Sulaiman R, Schetinin V, Sant P (2022) Review of machine learning approach on credit card fraud detection. Hum Centric Intell Syst 2(1):55–68. https://doi.org/10.1007/s44230-022-00004-0

Bishop CM (2006) Pattern recognition and machine learning (Information science and statistics). Springer-Verlag New York Inc

Borg A, Boldt M, Rosander O, Ahlstrand J (2021) E-mail classification with machine learning and word embeddings for improved customer support. Neural Comput Appl 33(6):1881–1902

Boschetti A, Massaron L (2018) Python data science essentials: a practitioner's guide covering essential data science principles, tools, and techniques, 3rd edn. Packt Publishing, Birmingham, UK

Boudabsa L, Filipović D (2022) Machine learning with kernels for portfolio valuation and risk management. Finan Stochast, 1–42

Cambria E, Das D, Bandyopadhyay S, Feraco A (eds) (2017) A practical guide to sentiment analysis, vol 5. Springer International Publishing, Cham

Chaboud AP, Chiquoine B, Hjalmarsson E, Vega C (2014) Rise of the machines: algorithmic trading in the foreign exchange market. J Financ 69(5):2045–2084. https://doi.org/10.1111/jofi.12186

Corea F (2019) How AI Is changing the insurance landscape. In: Applied artificial intelligence: where AI can be used in business (SpringerBriefs in complexity). Springer, Cham. https://doi.org/10.1007/978-3-319-77252-3_2

Cortes C, Vapnik VN (1995) Support-vector networks. Mach Learn 20:273–297

Cover T, Hart P (1967) Nearest neighbor pattern classification. IEEE Trans Inf Theory 13(1):21–27

Domowitz I, Yegerman H (2006) The cost of algorithmic trading: a first look at comparative performance. J Trading 1(1):33–42. https://doi.org/10.3905/jot.2006.609174

Eling M, Lehmann M (2018) The impact of digitalization on the insurance value chain and the insurability of risks. Geneva Pap Risk Insur Issues Pract 43:359–396

Farhadloo M, Rolland E (2016) Fundamentals of sentiment analysis and its applications. In: Pedrycz W, Chen SM (eds) Sentiment analysis and ontology engineering. studies in computational intelligence, vol 639. Springer, Cham, pp 1–24. https://doi.org/10.1007/978-3-319-30319-2_1

Fischer T, Krauss C (2018) Deep learning with long short-term memory networks for financial market predictions. Eur J Oper Res 270(2):654–669

Grigorev A (2021) Machine learning bookcamp: build a portfolio of real-life projects. Simon and Schuster

Gsell M (2008) Assessing the impact of algorithmic trading on markets: a simulation approach (No. 2008/49). CFS working paper. Goethe University Frankfurt, Center for Financial Studies (CFS), Frankfurt

Han J, Kamber M (2006) Data mining: concepts and techniques, 2nd edn. Morgan Kaufmann Publishers, San Francisco

Hendershott T, Jones CM, Menkveld AJ (2011) Does algorithmic trading improve liquidity? J Financ 66(1):1–33. https://doi.org/10.1111/j.1540-6261.2010.01624.x

Janiesch C, Zschech P, Heinrich K (2021) Machine learning and deep learning. Electron Mark 31(3):685–695

Jo T (2021) Machine learning foundations. In: Machine learning foundations. Springer Nature Switzerland AG. https://doi.org/10.1007/978-3-030-65900-4

Jordan MI, Mitchell TM (2015) Machine learning: trends, perspectives, and prospects. Sci 349(6245):255–260. https://doi.org/10.1126/science.aaa8415

Kissell R (2014) The science of algorithmic trading and portfolio management. Academic Press

LeCun Y, Bottou L, Bengio Y, Haffner P (1998) Gradient-based learning applied to document recognition. Proc IEEE 86(11):2278–2324

Mishra KN, Pandey SC (2021) Fraud prediction in smart societies using logistic regression and k-fold machine learning techniques. Wireless Pers Commun 119(2):1341–1367. https://doi.org/10.1007/s11277-021-08283-9

Mitchell TM (1997) Machine learning, vol 1, no 9. McGraw-Hill, New York

Mitchell TM (2006) The discipline of machine learning, vol 9. Pittsburgh, Carnegie Mellon University, School of Computer Science, Machine Learning Department, p 3

Mohammed M, Khan MB, Bashier EBM (2017) Machine learning: algorithms and applications. Crc Press, Taylor & Francis Group

Mohri M, Rostamizadeh A, Talwalkar A (2018) Foundations of machine learning. The MIT Press, 2nd edn. Cambridge, London, England

OECD (2019) Artificial intelligence in society. OECD Publishing, Paris. https://doi.org/10.1787/eedfee77-en

Paiva FD, Cardoso RTN, Hanaoka GP, Duarte WM (2019) Decision-making for financial trading: a fusion approach of machine learning and portfolio selection. Expert Syst Appl 115:635–655

Popovych B (2022) Application of AI in credit scoring modeling. Springer Gabler. https://doi.org/10.1007/978-3-658-40180-1

Prix J, Loistl O, Huetl M (2007) Algorithmic trading patterns in xetra orders. Eur J Finance 13(8):717–739. https://doi.org/10.1080/13518470701705538

Ramesh P, Wilpon JG (1992) Modeling state durations in hidden Markov models for automatic speech recognition. In: Acoustics, speech, and signal processing, IEEE international conference on, vol 1. IEEE computer society, pp 381–384

Sabbeh SF (2018) Machine-learning techniques for customer retention: a comparative study. Int J Adv Comput Sci Appl 9(2):273–281

Sahu MK (2019) Machine learning algorithms for automated underwriting in insurance: techniques, tools, and real-world applications. Distrib Learn Broad Appl Sci Res 5:286–326

Salvin P, Lepcha N (2022) Financial frauds and scams. In: Encyclopedia of gerontology and population aging. Springer International Publishing, Cham, pp 1855–1861

Samuel AL (1959) Some studies in machine learning using the game of checkers. IBM J Res Dev 3(3):210–229

Sarker IH (2021) Machine learning: algorithms, real-world applications and research directions. SN Comput Sci 2(3):160

Schapire RE (2013) Explaining adaboost. In: Empirical inference: festschrift in honor of vladimir N. Vapnik. Springer Berlin Heidelberg, Berlin, Heidelberg, pp 37–52

Shalev-Shwartz S, Ben-David S (2014) Understanding machine learning: from theory to algorithms. Cambridge University Press

Shen G, Tan Q, Zhang H, Zeng P, Xu J (2018) Deep learning with gated recurrent unit networks for financial sequence predictions. Procedia Comput Sci 131:895–903

Srinivas M, Sucharitha G, Matta A (eds) (2021) Machine learning algorithms and applications. Wiley

Ta VD, Liu CM, Tadesse DA (2020) Portfolio optimization-based stock prediction using long-short term memory network in quantitative trading. Appl Sci 10(2):437

Turing AM (2009) Computing machinery and intelligence. In: Epstein R, Roberts G, Beber G (eds) Parsing the turing test. Springer, Dordrecht. https://doi.org/10.1007/978-1-4020-6710-5_3

Venkatasubbu S, Perumalsamy J, Mohammed SB (2023) Machine learning models for life insurance risk assessment: techniques, applications, and case studies. J Artif Intell Res Appl 3(2):423–449

Wang W, Li W, Zhang N, Liu K (2020) Portfolio formation with preselection using deep learning from long-term financial data. Expert Syst Appl 143:113042

Watt J, Borhani R, Katsaggelos AK (2020) Machine learning refined: foundations, algorithms, and applications. Cambridge University Press

Williams G (2011) Data mining with Rattle and R: the art of excavating data for knowledge discovery. Springer Science & Business Media

Wu Y, Mao W, Feng J (2022) AI for online customer service: intent recognition and slot filling based on deep learning technology. Mob Netw Appl 27(6):2305–2317

Yang GJ, Zhou YM, Sun LL (2009) Enterprise financial early warning method based on benford-logistic model. J Quant Techn Econ 36(10):149–165

Digital Currencies and Financial Transformation

Bilal Bagis

Abstract The global economy is undergoing a fast digitalization process. Digitalization and interest in crypto technologies have accelerated, in particular post the Covid pandemic. Meanwhile, crypto and blockchain technologies represent a major design shift in terms of how the economies, financial transfers, payments, and even trade connectivity work. Cryptocurrencies and their technology have already started transforming the global financial and payment systems. Yet, central banks and other policy institutions also need to follow these trends closely. Digitalization itself, on the other hand, might be defined as the incorporation of digital instruments or new digital technologies into our lives. The role of ICTs (information and communication technologies) has substantially increased in economies, financial flows, payment systems, and even in social life. In that sense, digital currencies lead the contemporary major financial transformation era. They combine the privacy of physical central bank money and the convenience of electronic private bank money. The rise of e-commerce and digital payment systems, coupled with blockchain technology, has hence led to the emergence of digital currencies, much more efficient payment technologies, and further financial innovations. Financial systems will also be much simpler, as they get digitalized. Blockchain could potentially even replace internet technology. After all, the centralized internet structure and rising monopoly or oligopolistic power over certain services within modern internet technologies have already led to a major IT outage in 2024. In the short term, though, it will probably need to be used together with the internet.

Keywords Digitalization · Cryptocurrency · Digital currency · Blockchain

---

B. Bagis (✉)
Department of Economics and PERPAL Lab, Fatih Sultan Mehmet Vakıf University, Istanbul, Türkiye
e-mail: bilalbagis@yahoo.com

M. Gün and B. Kartal (eds.), *Machine Learning in Finance*, Contributions to Finance and Accounting, https://doi.org/10.1007/978-3-031-83266-6_2

1 Introduction

A global world and a digital world are two defining features of the new era ahead. Meanwhile, two recent major events (the COVID-19 pandemic of 2020—the Pandemic, and the Global Financial Crisis of 2008–2008 Crisis) have significantly improved global economic, financial, and social connectivity through more digitalization, alongside the developments in information technologies and communication instruments.

The power of digitalization and financial innovation lies particularly in improving the efficiency of international financial systems and multilateral payment systems. Yet, the digital transformation surely means more than that (WEF 2023). Out of over 8 billion total world population, more than 5 billion (more than 60%) already have social media accounts and internet access. Almost 6 billion of these people also have a unique phone number (around 70%) (IMD 2024).

Digitalization in finance is particularly fast-paced forward. For example, decentralized Finance (DeFi) is the new modern crypto-type transformation in financial markets and financial services (such as borrowing, lending, and trading). DeFi, eliminates banks, brokerage, and other intermediaries to avoid huge commissions and fees. Transactions are, instead, enabled via smart contracts. These smart contracts allow both sides of the trade together with a simultaneous exchange of money and assets.

Digital currencies, on the other hand, might even enable financial stability and help gain a bigger financial presence in the global economic and financial landscape. Accordingly, governments aim to utilize the technology behind digital currencies and have hence even come up with their sovereign digital currencies. Developing economies, particularly, are rising as leaders in this new era of digitalization and digital currencies.

The centralized information technologies and the internet structure as well as rising monopoly or oligopolistic power over certain IC services (within the modern internet technologies) have already led to a major IT outage in 2024. This major outage and the growing power concentration in conventional economic, financial, and even technological frameworks are noteworthy.

That is partly the reason the search for new ICTs has gained prominence in the past few decades. Blockchain, on the other hand, is a major alternative recording and accounting system. Hence is sometimes even called the internet of the 1970s. Furthermore, just as the internet technology has radically changed the technology of information flows; blockchain technology also seems to radically alter all types of transaction technologies and ways of doing business.

2 Digital Transformation

The epicenter of all these transformations, though, is digital transformation (Hatt et al. 2022). The need for a strong and secure digital infrastructure, digital and financial literacy, privacy protection measures, and inclusive digital public infrastructures (as in the Turkish e-Devlet) is eminent. Digitalization, therefore, requires a holistic approach even in the design stage.

Digitalization in public services and the need to increase financial efficiency is indisputable. The rise of e-commerce and digital payment technologies is a new reality. Even international organizations have increasingly been interested in this new trend. Accordingly, the theme of the 40th COMCEC meeting (in November 2024) and earlier OTS meetings was also a digital transformation of payment systems, for instance.

Digitalization of the global economy was always prevalent, considering modern electronic currencies and all other FinTechs (new apps, tokens, trading platforms, electronic payment systems such as Visa, Troy, etc.). FinTechs (financial technology), in particular, identify the type of financial businesses that utilize new technology to improve or mechanize the services they provide.

The rise of crypto and digital technologies, introduction of blockchain technologies, the 2008 Crisis and in particular the Pandemic have quickly accelerated this digital transformation trend. Though, reflections on behavioral economics, in terms of how (for instance) behavioral changes affect consumption culture, spending habits, and how it affects other aspects of economies, matter too.

Digital transformation has even turned into a necessity for companies, institutions, and even national economies at a macro level. They all need to follow the ongoing trend (footsteps of the private initiatives etc.). Their techno-optimism regarding the fast-approaching digitalization (too) is noteworthy.

For central banks and their national economies, for instance, looking forward, search for alternatives to the existing reserve currencies for trade (as in the BRICs or Brazil-Argentina examples), sanctions, and financial hegemony concerns; and in some cases, access to finance urges or ensuring accessibility for all, is further dominating this trend.

For instance, with the blockchain technology, trust, verification, certification, or the consensus mechanism is distributed within the networks, in a decentralized nature. Yet the new alternatives should also consider the risks towards financial stability, privacy, and monitoring.

Meanwhile, in addition to the past few centuries' primary general-purpose technologies (GPTs), including the blockchain revolution of the past few decades, there were mainly three major revolutions in the past 50-to-60 years (Bagis 2023):

- introduction of the internet and the new era of communication technologies,
- the early 2000s and the rise of social networks, smartphones, etc.,
- the pandemic and the third digital revolution (zoom, online shopping, etc.).

Table 1 Affordability of ICT services worldwide (price of a basket and GNP per capita ratio)

Regions	2021 (%)	2022 (%)
World	1.9	1.5
Africa	6.5	5.0
Americas	2.6	2.4
Arab region	1.2	1.0
Asia-Pacific	1.5	1.4
Commonwealth of Independent States	1.1	0.9
Europe	0.5	0.4

Source ITU (2022) andUnited Nations (2024)

Crypto technologies are a good example of this new era of digitalization. Bitcoin and other digital currencies have propagated a new era of digital payment networks, that are basically verified over a new technology called blockchain. Hence, the new trend of digital currencies and financial transformation.

With this discussion over digital currencies and financial transformation, the authors are emphasizing the process of redefining money and peer-to-peer payments, as well as other similar financial technologies. After all, the World is undergoing an era of major financial transformation (Itten et al. 2020). And digital currencies, blockchain technology are pioneers of this ongoing transition period.

Yet, the digitalization and structural transformation process should also be compatible with green transformation (Shahmohammadi et al. 2020). After all, as rightly pointed out in UNCTAD (2024), there is indeed an urgent need for green and sustainable digitalization strategies. In that sense, the ongoing global digital transformation trend is an opportunity not to be missed (Table 1 and ITU 2023).

The authors, therefore, focus on blockchain and financial transformation. More specifically on modern cryptocurrencies, the blockchain technology behind these new instruments, FinTechs, payment systems, and financial transformation. Though, surely, more focus is needed on P2P (peer-to-peer) or B2B (business-to-business) payment systems. That is the decentralized (distributed network system) payment or money transfer systems.

No doubt, financial transformation, should (first and foremost) mean more effective and efficient use of blockchain technology in finance. Just as in the case of FinTechs, new payment systems, CBDCs, cryptocurrencies, etc. Contributions of quantum computers and space technologies to automation in areas such as finance should also be noted. One needs also to contemplate whether there are other better future digital payment services or technologies (than just the CBDCs).

2.1 *Financial Digitalization*

When it comes to financial digitalization, the first thing that comes to mind is of course FinTechs, which include applications that provide banking or insurance services in a digital environment without any branches or intermediaries. In addition, new-generation payment systems and digital wallet popularity are also increasing with each passing day.

On the other hand, cryptocurrencies and the CBDCs (Central Bank Digital Currency) are among the game-changing elements in the field of finance or financial digitalization. Contributions of quantum computers and space technologies to automation, in finance particularly, are also worth emphasizing.

For instance, the use of quantum computers in optimization is becoming increasingly widespread. It is of great importance not to be late in investing in this new quantum computer technology too, which (for instance) might even allow very time-consuming and energy-consuming operations such as crypto mining to be carried out in a much shorter time.

Türkiye, for example, should probably aim to play a leading role, in areas where software is at the forefront, specifically in gaming, artificial intelligence, e-commerce, FinTechs, and other communication technologies. Meanwhile, as in the example of "a million software developers" project, public resources Türkiye has recently spent on digital transformation are at substantial amounts.

In particular, the country should not miss the opportunity to become a pioneer or leading country in blockchain technology and other technologies that start with it. Meanwhile, although blockchain technology is usually perceived as a new cryptocurrency investment area, by most people (even in Türkiye), it is indeed rather a technology that for example enables peer-to-peer encrypted transactions and can be used in many areas other than cryptocurrencies. In the meantime, since blockchain usage areas are just being built, it is of great importance for countries such as Türkiye to increase their effectiveness before the opportunity window narrows.

It is (indeed) a well-known fact how high the interest of the Turkish people is in crypto money and tokens. The grand opening of local crypto sales platforms has moved Türkiye from being just a consumer into the game of producers as well. However, beyond this, producing cryptocurrencies and tokens that will have a place on a global scale should also be among other duties and goals. The window of opportunity might even close soon (Robinson and Acemoglu 2012).

Meanwhile, one of the most crucial application areas of blockchain technology is smart contracts. Thanks to this application, even legal contracts can be signed with blockchain-encrypted algorithms and stored in a way that cannot be changed. This new field of application (still in its initial stages) can perhaps be considered as one of the new focus areas for Türkiye.

3 The Blockchain Technology

Leading GPTs (general-purpose technologies) that have since utterly transformed the world include steam engines, electric power, information technology (IT), internet technology, and artificial Intelligence (AI). And blockchain is just added to that list, Bagis (2023). Blockchain could potentially even replace internet technology. However, in the short term, blockchain technology will probably need to be used together with the internet.

Nonetheless, blockchain technology is at the forefront of digitalization and financial transformation. The technology is leading a major financial transformation. Blockchain is sometimes called the internet of the 1970s, as there were, indeed, several similar technologies as late as the 1970s. Nowadays, it pushes the frontiers of payment technologies. More specifically, it helps with;

- Traceability and provenance (SCM),
- Payment and settlements.

The Economist magazine once called blockchain technology the "trust machine" (Economist 2015). This is because blockchain technology primarily helps build trust. And it could, indeed, be extremely useful in building this much-needed trust, in particular for those (small players) that don't yet have a built-up trustworthiness, reputation, credit scores, etc.

Blockchain will, hence, be a trustworthy record keeper. It will benefit from the new science of keeping private information secret. Therefore, as rightly pointed out by Economist magazine, blockchain technology, the technology behind Bitcoin, is considered a new phenomenon that could potentially transform the way economies run (Economist 2015).

Blockchain technology itself was introduced with the digital currency innovation. Digital coins are minted using some cryptographic rules, set at the very beginning. These new-generation digital coins are then exchanged on decentralized wired (computer) networks and using digital wallets. And all the transactions are recorded on decentralized ledgers (called blockchains).

Blockchain enables information and transaction orders to be stored in blocks, which are connected in a sequence like rings of a chain (Bağış 2023b). It works on the principle that transactions are approved within a consensus mechanism.

Digital accounts (digital wallet) have two keys (passwords), one private and another public. This means a two-stage security system for transactions, to ensure digital security. Hence more security! Your private password is yours and remains yours only. The public password (though) is used by everyone, to confirm your transaction.

Thus, it has digital ledgers that (for instance) record monetary transfers on a network. However, since these ledgers are kept separately by everyone (as they are processed separately in everyone's digital ledgers), there is no chance for a single user to deceive all others. Meanwhile, the growth of the blockchain network will constantly increase security, thanks to the distributed data recording identity.

A critical warning in these early stages of development, progress, and finding its route, though, is that it is also important that the public authority does not prevent blockchain technology from finding its own identity. More specifically, settling and developing with too many regulations, inspections and bans would be detrimental. Cryptocurrency-like harsh and sometimes restrictive regulations should not apply to blockchain.

3.1 Benefits of Digitalization

The rise of e-commerce and digital payment technologies, coupled with blockchain technology, has led to the emergence of digital currencies, much more efficient payment technologies, and further financial innovations. The financial system will also be much simpler, as it gets digitalized.

Public procurement or public savings could also be strengthened with steps towards digitalization. For example, public stationery expenses, bills, and other physical expenses can be reduced; costs can be minimized with electronic notifications and e-correspondences.

Likewise, publications, reports, promotional documents, and other similar expenses could be reduced with electronic operations. Costs and expenses can also be reduced by switching to blockchain-based recording and transaction technologies. The Turkish Takasbank's blockchain-based transfer system is one good example (Anadolu Agency 2019).

Other key areas where blockchain is and could be effectively used, include but are not limited to;

- Transportation and supply chain management,
- Safety, privacy and compliance,
- Payments and settlements,
- Real estate records.

The technology could enable cheaper and safer public databases for land registries, art registries, or even notarization. It could be used for trading records at exchange markets or even for credit scores, by financial institutions. They are expected to replace public authorization mechanisms, banks, bureaucracy, clearing houses as well as other public central authorities (literally any institution in the trust or authorization business).

Digitalization could be particularly effective in payment services and is likely to be the future of payments. After all, according to the World Bank (2024) calculations, the average cost of remittances was almost 11% at the end of 2020. Blockchain-led digital technologies, however, are much cheaper, as they mean much lower average commissions (1 in 10 or 1 in 100). For example, the World Food Program (WFP) is effectively using these types of blockchain-based solutions to decrease its costs (by eliminating intermediaries).

Apart from this, other major advantages of Blockchain could be listed as follows:

- Free authentication (verification or confirmation, approval, and reconciliation) process,
- The speed and the possibility to make transactions in just a few seconds, or real-time instead of a few days,
- Much more efficiency with fewer people, resources, and effort spent,
- The probability of fraudulent transactions, corruption, and theft is much lower,
- The element of trust is at a higher level (lower counterparty risk),
- Elimination of the risks of losses in accounting books and the information in them,
- Security and clarity.

The most critical advantage, though, is being able to carry out an official transaction, money, and financial capital transfer for very small fees, without any intermediaries, borders, and other time restrictions. This is to the extent that, in an extreme case, domestic payment systems, banks, and even the central banks could be replaced by numerous ledgers around the world.

Blockchain technology and smart contracts eliminate the operational workload in traditional business models and is an alternative that provides transparency, speed, and trust to the parties. The need for acceleration, transparency, and digitalization of foreign trade transactions in the pandemic, when even document sending and control processes in international trade became difficult, is also particularly worth underlying here.

4 Digital Currencies

Digitalization is an inevitable process. The Economist magazine's January 1988 cover "Get Ready for a World Currency", calling for a gold coin or a cryptocurrency was a long-time phenomenon, asking to get ready for the Phoenix (Economist 1988). The magazine shared its prediction that, 30 years after the relevant issue (in 2018), this new world money (which it hypothetically called Phoenix) would replace other currencies and be used even instead of the national currencies.

Indeed, digital currencies are already a new monetary reality that could gradually replace conventional payment intermediaries such as physical currency, and even banks (to some extent), financial institutions, and other payment services. In the same Economist article, it is underlined that the money in question would inevitably be used first for the private sector and companies, and then for the states and other public authorities (Economist 1988).

Digital currencies could also enable easier access to popular green finance options. Using digital currencies in projects related to sustainability and green finance initiatives could help deal with financing issues. Yet they need to be regulated, the risks should be mitigated, and they should serve the sustainable development goals.

At first, cryptocurrencies revolutionized payment technologies by eliminating the need for a trustworthy intermediary such as a conventional commercial bank, and then the FinTechs (as in the earlier electronic currencies, mobile banking, trading platforms, etc.) emerged. They made direct peer-to-peer payments possible.

CBDC, on the other hand, is the latest addition, in the thousands of years of evolution of money (Adjepong-Boateng 2023; Eichengreen 2019; Brunnermeier et al. 2019). The evolution of money from barter to metal coinage or paper currency, and then to electronic currency, cryptocurrencies, and finally to digital currencies (as in the CBDCs) is surely worth a deeper analysis.

However, CBDCs are primarily central banks' reaction to the creation of cryptocurrencies, and their way of utilizing blockchain technology. Yet, they could also be drivers of further innovation. CBDC designs could push the limits, lead to new digital innovations in payments, and meet the future digital economy payment needs.

Blockchain-based currencies, and CBDCs in particular, also have a transformative capacity and aim to address contemporary issues related to digital payments. Blockchain technology has already catalyzed financial digitalization, helped introduce new payment technologies, and even the digitalization of money itself.

4.1 Central Bank Digital Currency

The contemporary rise of digital or cryptocurrencies has created new challenges for governments and policy institutions. Hence, a rush for new digital transformation and CBDCs. CBDC is, therefore, a new global phenomenon. A critical cornerstone in currency's long evolution process.

They revolutionize the financial system of a country, from top to toe. They are based on blockchain technology. They allow B2B or P2P transactions, excluding the necessity for a third intermediary. Yet, a fact: CBDC is still a process in the making (Singh et al. 2023).

CBDC is also an electronic form of fiat currency. It is a liability of the central bank or the government (legal tender). Most basically, they are digital banknotes. On the other hand, modern electronic currencies are backed by commercial banks (and some deposit guarantees, to some extent, by the governments or money at deposits). Cryptocurrencies, on the other hand, are tradable (just like commodities) and hence bear the risk as well as the return.

Conventional cryptocurrencies (such as Bitcoin or Ethereum) are verified by a network of multiple devices via a decentralized technology or a distributed ledger technology. CBDCs, on the other hand, are backed by central banks and hence could have a central hub (as in e-CNY).

Cryptocurrencies, though, are either mined or are a liability of a private issuer (if they are issued by a private entity). Their cousins' stablecoins are pegged to another currency, asset, or commodity. CBDCs, on the other hand, are the way public institutions are responding to all these (private) crypto technologies.

CBDCs aim to combine the privacy of physical central bank money and the convenience of electronic private bank money. They should (therefore) be more private than electronic private bank money. Indeed, in a token format, they can even be as private as cash or physical money.

CBDCs are a cornerstone in the evolution of money. Yet, they could also be key players in a safe, secure, and efficient digitalization process of the financial system and the modern economic system. CBDC is the future and has a critical place in future digitalization. It could contribute significant benefits to future financial infrastructure, and potentially even revolutionize the way trading and financial transactions are done.

Modern CBDCs are mainly intended for cross-border payments, and even for tokenization and interoperability. Almost 98% of the world economy and 134 sovereign countries (tripled since 2020) are working on ways to launch their sovereign digital currencies (Atlantic Council 2024).

As a matter of fact, all but 2% of the global economy and over 90% of the central banks have demonstrated an interest in the CBDCs. This figure is particularly valuable as just 3 years prior, almost a little over 30 of these economies were engaged with digital currencies (Bağış 2023a).

4.2 Benefits of Issuing CBDCs

The need to improve cross-border payment efficiency, accommodating digitalization, revivification of competition in the payment systems, the tokenization trends, and revivifying weak financial inclusion (as well as the institutional adjustments to the new era) are all raising interest in the CBDCs.

The primary value-added of CBDCs, though, is rising efficiency, increasing competition (in payment technologies), and the stimulation of innovation. Thus, a balance between competition and cooperation is needed, in payment systems (both at the domestic and at global level).

Payment efficiency and innovations are other two key strengths of CBDCs (in addition to the improvements in monetary policy and central banking), Bagis (2022). And public endorsement for and government implementations of CBDCs have also recently gained momentum.

Most central banks across the world are currently actively exploring CBDC (Bagis 2022) due to the ongoing;

- relatively high inflation rates,
- rising income inequality,
- the need for technological literacy,
- information literacy,
- and low financial inclusion.

All these factors influence public's perception regarding the CBDCs too. Further potential critical uses of CBDCs may also be listed as follows: More effective and sovereign monetary policy, increasing financial stability, enabling efficiency in financial-commercial transactions, efficient global payment systems, remittance transfers and financial transaction flows.

Meanwhile, retail CBDCs are used by individuals for payments. They are expected to enhance payment efficiency and financial inclusion. Particularly in the developing world, more inclusive financial systems and better access to financial services (especially) for unbanked citizens are still in order. Hence, digitalization and retail CBDCs would give anyone universal access to a digital sovereign currency.

Digitalization, innovation, data-driven approach to efficiency, productivity, austerity and even saving is still a long way forward. Yet, these processes could also improve ratings and access to lending. For instance, better data management and access to bigger data could fasten credit profile creation, help get information on credit risks much faster, and even help individuals or SMEs build a credit rating much earlier to access better funding.

Wholesale CBDC, on the other hand, is used by institutions for payment settlements or swaps. The modern reserve systems (reserve accounts with the central bank) between central banks and commercial banks already resemble these wholesale CBDCs. Therefore, large financial institutions already have access to digital central bank currency. Hence, for some, these wholesale CBDCs even already exist.

Particularly in advanced world economies, sustainability and strength of the financial system could surely be improved with the wholesale CBDCs. Yet, the reality is that since CBDCs are expected to improve financial efficiency, developing economies are almost twice as interested in CBDCs, and usually are at much more advanced stages of their CBDC projects. Meanwhile, in more advanced world economies such as Denmark, Japan, and Australia, payment systems are considered advanced. Thus, no urgent need is deemed for CBDCs.

Decreasing the costs of financial transactions is another reason for rising interest. The global average cost of sending money abroad (for wholesale, retail purchases, or remittances) was 6.25%, in 2023, according to the World Bank (World Bank 2024). The current maximum is over 50% (from Türkiye to Bulgaria, for example, the cost is as high as %52).

Cashless transactions would also help central banks improve their control over the financial system and payment systems. Illicit financial transactions and money laundering would also be limited or controlled.

More specifically, cryptocurrencies have also been weakening public authority over monetary and fiscal policy. With the CBDCs, public institutions' authority over money supply and new means to implement fiscal policy would be regained. Costs related to physical currencies and cash management (too) would be eliminated, and more transparency related to transactions (mitigated underground activities) would allow governments to collect more taxes.

CBDCs might help with managing expectations, better management of monetary transfers, controlling the side effects of policies, and regenerating the currency or liquidity flows across the economy (aiming for targeted money flows).

4.3 CBDC Design

CBDCs' or digital currency's design is yet to be determined. And there is probably no one-size-fits-all solution to introduce them as well. As rightly pointed out by Georgieva (2022), there is no universally accepted approach to the CBDCs. Country-specific factors may also lead to varying needs or design features.

Meanwhile, the BIS, the Fed, and 6 other major central banks have already allied to determine the fundamentals of designing a CBDC. Accordingly, the CBDCs should,

- Not harm commercial bank intermediation or harm their deposit or loan flows,
- Negative impacts over monetary policy and the monetary transmission are expected to be limited or restricted,
- These domestic payment system improvements are also expected to have international impacts, as in the cross-border currency competition,
- Be designed as a new form of money that enhances digital payments.

On the other hand, five primary factors that directly influence CBDCs' usage and design may be listed as:

- Cross-border payments,
- Global accessibility,
- Privacy assurance,
- Stable and strong operations,
- Cash and real-time payment features.

Thus, CBDC is a new technology that can be directed and implemented in almost any way. Therefore, the design stage matters a lot. The way CBDCs will be designed is still a process in the making, though. Active discussions, current research, and pilot projects are all very critical in this design process. One thing is for sure, though, that universal and equal access for all citizens, is a must.

Yet, in the meantime, at this very early design stage, criticism is essential to have a useful, more democratic CBDC. Criticism has so far helped many world economies redesign their processes and their offers. Essentially, CBDCs should be strengthened with more democratic values rather than more government control over individual freedom. OECD's "CBDCs with democratic values" initiative is one good example at this very point.

Meanwhile, international cooperation to build a consensus framework (the Global Digital Compact) to govern digital technology and AI, as well as providing a secure and free digital future was accepted by the UN in September 2024, United Nations (2024).

The IMD World Digital Competitiveness (WDC) ranking for the adoption of digital technologies in public services, governance, business life, and society also demonstrates cross-country differences in digital technology adoptions (IMD 2024). More specifically, the report points to huge developing and developed economic disparities in digital transformations (Zallio and Clarkson 2022).

Furthermore, OECD's Digital Government Index (DGI) of public institutional efforts to transform and digitalize (as well as incorporate AI into its services) the ability of governments to provide public services and operate efficiently and effectively (even during the crisis) points to strong digital public foundations in most developed economies, OECD (2024). Türkiye's DGI ranking is close to the OECD average.

Additionally, as a good example of effective digital design, cross-border CBDCs could be the future. They could help with better coordination of world economies, compared to the other options. Cross-border CBDCs, in a way, are about giving access to the other country's banks to the home country's wholesale CBDC (central banks allowing access to each other's wholesale CBDCs).

It could provide the world economies with strategic, geopolitical independence, looking forward. Cross-border CBDCs could help democratic economies decrease their dependence on unfriendly nations, and certain private companies and their vulnerability to similar shocks. Even the BRICS economies are currently working on ways to build a blockchain-backed digital payment system. It could help avoid too expensive intermediaries and SWIFT-type intermediations.

On the other hand, a cash-type CBDC provides anonymity and privacy, while a deposit-type CBDC may lead to safer or faster transactions. Account-based (deposit type) CBDC provides traceability and extended control over the financial system, while token-based ones are anonymous and provide privacy.

5 Issues to Deal with

In the meantime, there are also numerous crypto-related issues to deal with. The problems imposed by decentralized financial technologies and digital currencies are worth contemplating. As an example, cryptocurrencies need to find alternative ways to minimize their huge energy consumption to run their ledgers continuously (Table 2 and Farfan and Lohrmann 2023). But this is (surely) not the only issue to contemplate.

Most crypto assets appear to have no inherent problem to solve, to make life easier. Therefore, their real values are always zero. They will only have value as long as individuals themselves accept them (the conventional fiat currencies are—at least—backed by governments). Meanwhile, all existing cryptocurrencies appear to be experimental. Their values can go up or down at any time. The only valuable fact is the blockchain technology behind them.

Furthermore, in their current state, they are more likely to create new problems and grievances than provide any solutions. (While they should be environmentally friendly) With their high energy requirements, they currently cause more harm to the environment. Both mining and proofing or recording the transactions on blocks create costs. This is, indeed, contrary to the green transformation and digital new structural transformation arguments, as in Hynes (2022), Muench et al. (2022), and Siragusa and Tumino (2022).

Table 2 Bitcoin's energy consumption (2010–2023), in TWh

Years	Energy consumption (TWh)
2023	121
2022	108
2021	105
2020	69
2019	57
2018	45
2017	14
2016	5.5
2015	3.5
2014	4.8
2013	1
2012	0.1
2011	0.1

Source UNCTAD (2024)

They also distort income equality to the detriment of those who do not know the issue (they have, somewhat, turned into a tool of smearing and unfair wealth distribution).

There are also concerns that the CBDCs could indeed;

- Increase power concentration over the central banks and lead to an institution-wise political hegemony,
- Decrease financial freedom and the ability to make economic or financial choices,
- Allowing any institution to monitor and surveillance of even private financial affairs.
- Potentially putting limits on the amount you can spend,
- Even putting an expiration date on currency to ensure they are spent,
- Taxing money holding,
- The ability to freeze financial assets, wealth, or liquidity of citizens at any point,
- Financial stability concerns as in a potential bank disintermediation.

Another critical weakness of digital currencies (and of these types of financial technologies, in general) is that, as the banking panic of 2023 in the US has demonstrated, new financial innovations have indeed facilitated much quicker and heavier bank runs.

CBDCs should be designed in such a way that privacy and anonymity are protected to the extent that illegal activities such as tax evasion, financing terrorism, money laundering, and other criminal actions are prevented. Some kind of managed anonymity might be needed. Potential negative implications on the workplace should be dealt with, while the process is effectively utilized to improve technology leadership, looking forward (Bagis 2024).

The potential impacts of CBDCs on financial stability and efficiency of monetary policy transmission are other issues to contemplate. CBDCs should not (and probably will not) replace cash or other fast payment systems. It is still not clear (though) if it will mean decentralized finance, financial decoupling, and a limited role for financial sanctions.

5.1 Digitalization and Green Finance

Another crucial aspect of the ongoing digital transformation era is its transition towards green finance. Green finance would mean environmentally friendly, respectful loans and financing options. Digital finance, digital economies, green finance, and rising efficiencies are also mostly interrelated. Monetary policy efficiency, access to green finance options, and digital currencies also work hand in hand.

Compatibility with the green transformation matters. Green banking and green finance applications (as in developing green banking applications within the scope of compliance with the European Green Deal) should also be considered, looking forward. Green bonds or green sukuk have recently gained popularity even in the richer Western world.

Financial technology companies (FinTechs), new and innovative initiatives (startups), new data centers, agricultural technologies, and new energy investments (green investments, particularly) that emerge as alternatives to the traditional technologies and their innovative finances are evident, in that regard.

Innovations such as cryptocurrencies are surely gaining more prominence in parallel with the green transformation. However, innovative cryptocurrencies such as Bitcoin, in their current formation, are (unfortunately) not environmentally friendly as (still) too much energy is used in mining (see Table 2 and CODES 2022).

6 Conclusion

Digitalization is particularly fast-paced forward. Yet, there is still too much room for innovation and development, when it comes to digital currencies. For instance, AI could be more effectively used to improve and fasten credit rationing, accounting, and credit score evaluations. It could also help personalize financial services, in particular for those with limited financial literacy.

The trust issue, in this new era of blockchain-based digitalization, is provided by cryptographic and mathematical formulas. In the meantime, there are too many uncertainties around their usage. The public sector, though, should get ready for this innovation and other related payment technologies. Regulations are already underway and helping shape this instrument.

Blockchain could certainly be one of the strategic areas where smart power and central position can be strengthened for new global players such as Türkiye. With significant investment in this new technology, countries such as Türkiye (which has an important potential in terms of strategic and geographical position, as well as its young population) can capture the privileged position that India has in software or Silicon Valley in computer technology etc.

One critical warning (in these early stages of development, progress, and finding its route), though, is that it is also important that the public authority does not prevent blockchain technology from finding its own identity; settling and developing with too many regulations, inspections and bans. Cryptocurrency-like harsh and sometimes restrictive regulations should probably not apply to blockchain and digital transformation.

References

Adjepong-Boateng K (2023) Recent books on the history of money. Hist J 66(2):480–495. https://doi.org/10.1017/S0018246X22000346

Anadolu Agency (2019) Takasbank'ın blokzincir tabanlı transfer platformu hizmete girdi, 30 Dec 2019. https://www.aa.com.tr/tr/ekonomi/takasbankin-blokzincir-tabanli-transfer-platformu-hizmete-girdi/1687596

Atlantic Council (2024) Central bank digital currency tracker, 20 Sept 2024. https://www.atlanticcouncil.org/cbdctracker/

Bagis B (2022) Digital currencies and monetary policy in the new era. Insight Turk 24(3):189–211. https://doi.org/10.25253/99.2022243.12

Bagis B (2023) Could blockchain be a real game changer? Politics Today, 26 Jan 2023. https://politicstoday.org/could-blockchain-be-a-real-game-changer/

Bağış B (2023a) The rise of central bank digital currency. Daily Sabah, 8 Aug 2023. https://www.dailysabah.com/opinion/op-ed/the-rise-of-central-bank-digital-currency

Bağış B (2023b) Review of the rise of blockchains: disrupting economies and transforming societies, by N Kshetri, Edward Elgar. Insight Turk 25(1):276–278

Bagis B (2024) Review of leading the digital workforce: IT leadership peak performance and agility, by JW Brown, CRC Press, Taylor and Francis Group

Brunnermeier MK, James H, Landau JP (2019) The digitalization of money. National Bureau of Economic Research, No. w26300

CODES (2022) Action plan for a sustainable planet in the digital age. Coalition for Digital Environmental Sustainability, Geneva

Eichengreen B (2019) From commodity to fiat and now to crypto: what does history tell us? National Bureau of Economic Research, No. w25426

Economist (1988) Get ready for a world currency, 9 Jan 1988

Economist (2015) The trust machine: how the technology behind bitcoin could change the world, 31 Oct 2015. https://www.economist.com/weeklyedition/2015-10-31

Farfan J, Lohrmann A (2023) Gone with the clouds: estimating the electricity and water footprint of digital data services in Europe. Energy Convers Manage 290:117225

Georgieva K (2022) The future of money: gearing up for central bank digital currency. International Monetary Fund (IMF)

Hatt T, Iacopino P, Jarich P (2022) Radar: digital transformation in a post-pandemic future. GSMA intelligence. Global System for Mobile Communications Association, London

Hynes M (2022) Virtual consumption: a review of digitalization's "green" credentials. Front Sustain 3:969329

IMD (2024) World digital competitiveness ranking 2023. International Institute for Management Development. https://www.imd.org/centers/wcc/world-competitiveness-center/

ITU (2022) Affordability of ICT services. International Telecommunication Union. www.itu.int/itu-d/reports/statistics/2022/11/24/ff22-affordability-of-ict-services

ITU (2023) Measuring digital development: facts and figures 2023. International Telecommunication Union: Development Sector, 27 Nov

Itten R, Hischier R, Andrae ASG, Bieser JCT, Cabernard L, Falke A, Ferreboeuf H, Hilty LM, Keller RL, Lees-Perasso E, Preist C, Stucki M (2020) Digital transformation—life cycle assessment of digital services, multifunctional devices and cloud computing. Int J Life Cycle Assess 25:2093–2098

Muench S, Stoermer E, Jensen K, Asikainen T, Salvi M, Scapolo F (2022) Towards a green and digital future: key requirements for successful twin transitions in the European Union. Publications Office of the European Union. JRC 129319. Luxembourg

OECD (2024) 2023 OECD digital government index: results and key findings. OECD Public Governance Policy Papers, No. 44. OECD Publishing, Paris. https://doi.org/10.1787/1a89ed5e-en

Robinson JA, Acemoglu D (2012) Why nations fail: the origins of power, prosperity and poverty. Profile, London, pp 45–47

Shahmohammadi S, Steinmann ZJN, Tambjerg L, van Loon P, King JMH, Huijbregts MAJ (2020) Comparative greenhouse gas footprinting of online versus traditional shopping for fast-moving consumer goods: a stochastic approach. Environ Sci Technol 54(6):3499–3509

Singh S, Gupta S, Kaur S (2023) The quest for CBDC: indentifying and prioritising the motivations for launching central bank digital currencies in emerging countries. Qual Quant 57:4493–4508. https://doi.org/10.1007/s11135-022-01558-8

Siragusa C, Tumino A (2022) E-grocery: Comparing the environmental impacts of the online and offline purchasing processes. Int J Log Res Appl 25(8):1164–1190

UNCTAD (2024) The digital economy report 2024. The United Nations. https://unctad.org/system/files/official-document/der2024_en.pdf

United Nations (2024) Global digital compact. https://www.un.org/techenvoy/global-digital-compact

WEF (2023) Global risks report 2023 (Insight Report), 18th edn. World Economic Forum, Geneva

World Bank (2024) Remittance prices worldwide, 30 Sept. https://remittanceprices.worldbank.org

Zallio M, Clarkson PJ (2022) Designing the metaverse: a study on inclusion, diversity, equity, accessibility and safety for digital immersive environments. Telematics Inform 75:101–109

A Hybrid ARIMA-LSTM/GRU Model for Forecasting Monthly Trends in Turkey's Gold and Currency Markets with a Macro-Economic Data-Driven Approach

Mehmet Fatih Sert

Abstract This research introduces a methodology that presents a hybrid model that integrates ARIMA with LSTM/GRU architectures to forecast monthly trends in the Turkish gold and foreign exchange markets. The volatility inherent in the economic frameworks of emerging economies such as Turkey has become increasingly evident, especially following the sudden fluctuations in exchange rates and gold prices that started in 2018. In this context, traditional time series models struggle to effectively model complex economic behavior. To address these challenges, this study formulates a hybrid model that combines the advantages of ARIMA with the nonlinear learning capabilities of the LSTM and GRU deep learning frameworks. The dataset used covers a period of 24.5 years from 2000 to 2024. The model includes both Turkey-specific and global macroeconomic factors. To improve the forecasting accuracy of the model a Walk-Forward validation approach was used to continuously improve the model with each successive observation. The findings show that the proposed model achieves high success in both short-term and long-term forecasting of gold and exchange rates and effectively adapts to sudden market changes.

Keywords ARIMA-LSTM/GRU integration · Hybrid models · Gold forecasting · Currency trends · Machine learning

1 Introduction

The economies of emerging economies such as Turkey which exhibit noteworthy fluctuations and large changes in economic indicators when analyzed in the long run, involve significant vagueness for investors and policymakers. Specifically, in recent years, exchange rates, gold prices, and interest rates have seen sudden and unexpected changes. The volatility of local currencies against exchange rates, which has

M. F. Sert (✉)
The Faculty of Economics and Administrative Sciences, Gaziantep University, Gaziantep, Türkiye
e-mail: mfsert@gantep.edu.tr

M. Gün and B. Kartal (eds.), *Machine Learning in Finance*, Contributions to Finance and Accounting, https://doi.org/10.1007/978-3-031-83266-6_3

accelerated since 2018, has further increased the difficulty of forecasting the future of the economy and currency prices in the long term. Along with the pandemic, geopolitical, and conjectural developments, there has been vagueness in many countries and global economic indicators, making it difficult to forecast economic variables on a global scale. However, for countries with fragile economic structures such as Turkey, this situation poses a particular challenge for forecasting models.

Variables such as exchange rates and gold prices are heavily influenced by both global and local factors. In Turkey, in addition to global factors, indicators such as inflation rates, policy interest rates, and industrial production index change unpredictably. The high inflation, currency crises, and political fluctuations experienced in Turkey in certain years reduce the accuracy of forecasting models on foreign exchange and gold. In these cases, the predictive power of existing models is weak, and new approaches are needed for more accurate forecasts.

The success of traditional methods in modeling sudden changes in economic variables is limited (Jawadi 2012). Time series models such as ARIMA, which can be effective in predictable situations such as long-term trends and seasonality, may not be able to forecast successfully against large unpredictable fluctuations. On the other hand, machine learning and deep learning algorithms are gaining attention in terms of data processing and achieving significant results. However, the ability of these models to generalize in the long run may be limited, as they carry the risk of overfitting in long-term economic forecasts when used alone.

This research aims to develop a hybrid model to improve the efficiency of forecasting economic variables. A hybrid model is introduced that combines the strengths of the ARIMA model with the improved forecasting performance of deep learning architectures such as LSTM (Long Short Term Memory) and GRU (Gated Recurrent Unit). The main objective of the hybrid model introduced in this paper is to exploit the unique adaptive capabilities of deep learning algorithms to improve forecasting accuracy. At the same time, it incorporates the basic forecasting principles of traditional time series methods. Especially in economies prone to sudden fluctuations, such as Turkey, this model can provide a significant advantage in increasing forecasting capacity. For example, the rapid increases in inflation and sudden changes in exchange rates in 2018 and 2021 showed that classical modeling was insufficient in some cases and demonstrated the need for new techniques. The hybrid model can handle such volatility more flexibly and can take into account unexpected changes in economic indicators. Therefore, in our study, we combine ARIMA and LSTM/GRU methods to forecast gold prices and exchange rates in Turkey more efficiently.

Financial time series forecasting is an important research topic in economics and finance because macroeconomic indicators such as exchange rates, gold prices, and interest rates have an impact on both investment strategies and policy decisions. (Gubler 2012) There is a large literature on forecasting these variables. The ARIMA model, which is considered a classical method, is a frequently preferred approach in time series analysis (Yadav et al. 2024) and has yielded effective results in many studies on fluctuations in the Turkish economy. For example, Şahin (2023), Mashadihasanli (2022), Ashour and Al-Dahhan (2020), and Özmen and Şanlı (2017) obtained successful forecasting results.

Today, machine learning and deep learning techniques are gaining increasing importance in financial forecasting (Olubusola et al. 2024). Machine learning (ML) based approaches are particularly good at uncovering hidden patterns in large data sets. This makes them one step ahead (Jang and Lee 2019). The most prominent ML-based time series forecasting methods are Long Short Term Memory (LSTM) and Gated Recurrent Unit (GRU). These models can provide high accuracy when identifying complex relationships and trends (Ghadimpour and Ebrahimi 2022). However, the application of these models to large data sets poses some problems. Because long training time and high computational cost are required. Nevertheless, there is a large literature on trend forecasting, especially in gold and foreign exchange markets. (e.g., Ismail et al. 2009; Manoj and Suresh 2019; Suranart et al. 2014; Tunçsiper 2023; Plakandaras et al. 2015; Pierdzioch et al. 2015; ul Sami and Junejo 2017; Chandar et al. 2016; Wagh et al. 2022; Jabeur et al. 2024; Mithu et al. 2021; Livieris et al. 2020; Bingöl et al. 2020; Sudimanto et al. 2021; Kilimci 2022; Cordero et al. 2020; Manjula and Karthikeyan 2019; Cohen and Aiche 2023; Yamaçli and Yamaçli 2023; Küçük 2023; Igboanusi 2024; Landge et al. 2024).

2 Data

The time range of the data set was tried to be kept as wide as possible. Because it was desired to examine the changes in Turkey's economic indicators from a long-term perspective by using both local and global macroeconomic variables. The data set used in this study covers 24.5 years (January 2000–June 2024). This extensive period covers both globally and locally significant economic events, thus allowing the model to be trained for different events. For example, important periods such as the 2001 and 2008 economic crises in Turkey, the currency and inflation increases between 2018–23, the 2008 financial crisis globally, and the repercussions of the COVID-19 pandemic on the economy increase the model's ability to forecast changes in foreign exchange and gold prices.

Another dimension other than the duration of the data set is the number of variables included in the data set. By combining both Turkey's economic variables and global macroeconomic variables, a comprehensive data set was created with twenty variables. In this sense, the variables used in this study are compiled under two main classifications: Turkey-specific variables and global variables. The optimum number of variables is included in the model. This is because a small number of variables may not lead to the desired success in forecasting accuracy, while a large number of variables would create an excessive CPU performance and increase the time cost of the model at the point of producing output. Therefore, those of the main economic variables for which complete monthly data were available during the specified period were included in the model.

These variables are specific to Turkey:

- The inflation rate observed in Turkey (Trading Economics, n.d.),

- The policy interest rate set by the Central Bank of the Republic of Turkey (Federal Reserve Bank of St. Louis, n.d.)
- Current Account Deficit Data (Trading Economics, n.d.),
- Industrial Production Index (Turkish Statistical Institute, n.d.),
- BIST 100 Index (Investing.com, n.d.),
- BIST 100 Trading Volume (Investing.com, n.d.),
- Gold Prices (Monthly Average Bullion Gold Selling Price) (Central Bank of the Republic of Turkey, n.d.),
- USD/TRY Exchange Rate (Central Bank of the Republic of Turkey, n.d.),
- EUR/TRY Exchange Rate (Central Bank of the Republic of Turkey, n.d.).

The following data were collected as global variables:

- US Inflation Rate (Federal Reserve Bank of St. Louis, n.d.),
- Interest rates set by the US Federal Reserve (Federal Reserve Bank of St. Louis, n.d.),
- US 10-year bond rate (Federal Reserve Bank of St. Louis, n.d.),
- Brent Oil Prices (Energy Information Administration, n.d.),
- US Volatility Index, (Federal Reserve Bank of St. Louis, n.d.)
- US Industrial Production Index (Federal Reserve Bank of St. Louis, n.d.),
- Dollar Index (DXY) (Investing.com, n.d.),
- Eurozone Inflation Rate (2015 Based) (Eurostat, n.d.),
- Eurozone Industrial Production Index (OECD, n.d.),
- Eurozone Unemployment Rate (OECD, n.d.),
- Interest rates set by the European Central Bank (European Central Bank, n.d.).

3 Methodology

3.1 Arima

The first model used in the study is the ARIMA (AutoRegressive Integrated Moving Average) model, which is frequently preferred in the forecasting of financial time series. ARIMA is one of the basic econometric methods used to forecast future values of time series data by utilizing historical observations. The main purpose of ARIMA is to effectively capture linear relationships within time series and make forecasts about future values.

The ARIMA model has three main components:

AutoRegressive (AR) Component: This component allows the use of past values from the time series to forecast the current value. AR is related to the autocorrelation that exists between previous observations. The value of a series at time t is expressed as a linear combination of previous values. For example, the current value of the gold price is forecasted as a function of its prices over several previous periods.

Integrated (I): Integration is used to stabilize the time series by removing trends or seasonal fluctuations in the data. This ensures that the mean of the series remains constant over time.

Moving Average (MA) Component: It reveals the effect of the error terms of the previous values in the time series on the current value. By including the MA, it allows the model to forecast the present value not only with past values but also by taking into account past forecast errors.

The ARIMA model is expressed through parameters (p, d, q) that collectively integrate these three components:

p: Indicates the amount of autoregressive terms (AR component).

d: Relates to the number of differentiation procedures (the number of times the series must be differenced to make it stationary) (I component).

q: Indicates the number of moving average terms (MA component).

The ARIMA model shows considerable success in forecasting as it forecasts future values by capturing the underlying trends and seasonal patterns in the data. (Giovani et al. 2022) Its interpretation is simple due to its basic logic. It is especially successful in time series with trends and seasonality. However, the ARIMA model cannot model non-linear relationships. This reduces the accuracy of the model, especially in complex and volatile market scenarios. It also generally fails to excel in the presence of sudden shocks or significant fluctuations in time series. Moreover, ARIMA is only a univariate model. Naturally, it forecasts a single variable and cannot model the interrelationships among multiple variables. Therefore, it fails to take into account the effects of other macroeconomic indicators.

3.2 LSTM (Long Short-Term Memory) Model

LSTM is a specialized variation of the recurrent neural network (RNN) architecture, specifically designed to capture long-term dependencies by leveraging historical data within sequential datasets, including time series. Developed to solve the problem of "information loss over time" found in traditional RNNs, the model can effectively retain important historical information within the model. The key features of the LSTM model are:

Memory Cells: LSTM can capture both short-term and long-term addictions. Memory cells facilitate the retention of important information over long periods. As a result, it enables the identification of trends and seasonality in time series data.

Input, Output, Gates Unit: The LSTM model has three basic gate mechanisms. These are forgetting input and output gates. The "forget gate" is used to delete information, the "input gate" is used to store new information, and the "exit gate" is used to control what is removed from memory. These gate mechanisms eliminate irrelevant data in the LSTM model while important information is preserved. Thus, the accuracy of long-term predictions increases.

LSTM model can effectively detect complex patterns in time series data thanks to its ability to learn both long-term and short-term dependencies. It can capture trends and fluctuations. Therefore, it is a particularly useful tool for economic predictions. Because both trends and sudden changes are usually included in economic predictions. On the other hand, LSTM models may also face some difficulties such as long training times and overfitting sensitivity in large datasets.

3.3 GRU (Gated Recurrent Unit) Model

This model performs similarly to LSTM with a simpler and more efficient structure. GRU is an improved version of the RNN model. It is quite successful in determining long-term dependencies. Compared to LSTM, the training time is shorter because it has fewer parameters. The main difference of GRU is that it uses simpler update and reset gates instead of the input, forget, and output gates in LSTM. While the update gate provides the integration of new information, the reset gate determines how much of the previous data will be kept. (Dey and Salem 2017).

Since the GRU model has fewer gates and parameters, it is more efficient than LSTM in terms of both training time and computational cost (Elsayed et al. 2018). This provides a great advantage especially when working with large datasets. However, this simpler structure of GRU may not perform as well as LSTM in the face of more complex problems.

The greatest strength of these models is their ability to capture long-term dependencies and complex patterns (non-linear relationships) in the data (Ghadimpour and Ebrahimi 2022). Therefore, they are particularly successful in variables such as gold prices and exchange rates. Both models include gating mechanisms that facilitate adaptive management of data inputs, historical data, and forecast errors. This allows the modeling of complex relationships in the data just mentioned. Compared to classical time series models, these models are more successful in this regard. As a result, they provide a detailed analysis of complex relationships and sudden changes that frequently occur in financial markets. When choosing between LSTM and GRU, it is important to consider the advantages and disadvantages of each model. In cases where fast prediction and lower computational costs are required, GRU may be more appropriate. In cases where high accuracy is required and long-term dependencies are present, LSTM may be more appropriate.

3.4 Walk-Forward Validation

In traditional and static forecasting methods, the forecast result is obtained with a fixed training set and a one-time training phase. Financial time series are usually based on outdated data. They can also give inadequate results because they cannot adapt to changing conditions and economic indicators. The forward validation method developed to solve this problem involves incrementally updating the model over time (Kumar et al. 2023).

In this framework, the model is retrained with each new observation so that it works with the most recent data at each prediction step. In summary, at each stage, the model is supplemented with new data and tasked with forecasting the next period. Each time the model makes a forecast, new data is added and the model is retrained again to make a forecast for the next step. This iterative cycle continues until the entire dataset is processed (Pardo 1992). To explain the method in more detail, step by step, follow these steps: Initially, the model is trained on a subset of the available dataset, and forecasting is generated from the remainder. After this training phase, a one-step forecast is obtained from the test set. An example is the forecasting of the gold price for the following month. Once the actual value corresponding to the forecasting made by the model is obtained, i.e. as new data arrives, this information is integrated into the training set, thus extending the training data set. The model is then retrained using the newly included data and then forecasts are generated for the next period. This process is repeated at each stage and results in continuous updating of the model. The model continues to update with each forecast and this process continues until all observations in the dataset are accounted for.

This method improves the accuracy of forecasts by providing the model with the most recent data at each forecasting stage. This method provides a great advantage, especially for dynamic data sets such as those in financial markets. Allowing the model to be updated with each new observation allows the forecast to respond quickly to changes in market conditions and more effectively capture sudden fluctuations. It also reduces the risk of overfitting, as the model does not rely solely on historical data and is constantly refreshed with current information. However, applying this method requires retraining the model at each prediction step, which can lead to significant computational costs, especially for large data sets and complex models. In the case of deep learning models, this process requires both high computing power and time. Since retraining is done with each prediction, the complexity of the model can increase and in some cases, there is a risk of overestimating or making incorrect predictions. The prediction success of the model depends on the accuracy of the new data added. If there are errors or omissions in the new data, the forecasting performance may be negatively affected.

4 Model Implementation

4.1 Data Preparation and Scaling

After the data were collected, aggregated and a data warehouse was created, the data were subjected to a series of pre-processing steps to make them suitable for time series analyses. In each algorithm, the variables to be forecasted as dependent variables (Monthly average gold bullion sales price, USD/TRY, EUR/TRY Exchange Rate) were scaled by Min-Max, and the data were transformed into the range of 0 to

1, i.e. normalized. This scaling technique was performed to increase the speed and accuracy of learning processes in deep learning models.

In addition, standardization was applied to the other variables in each model. Accordingly, each variable was scaled so that its mean was zero and its standard deviation was one. In this way, all variables of the model can function effectively on a uniform scale. The reason for using different scaling techniques in the dependent and independent variables here is that as a result of the tests, it was understood that the model works most efficiently in this way. In addition, although the data set used in the study contains data until June 2024, the model was built with data until March 2024. Because the last 3 months' data were used to evaluate the over-forecasting success of the model on future values.

4.2 Delay Values

To increase the predictive power of the model, values corresponding to one (t − 1) and two (t − 2) time step lags of each variable are included as new variables in the model. In this way, the ability of the model to learn from past data is increased. The logic of including lag values in the time series data lies in the fact that the past states of the data are also used as data for forecasting purposes. With the inclusion of t − 1 and t − 2 lag values of the 20 variables in the model, a total of 60 dependent variables were included in the model.

4.3 ARIMA Model

In this study, the ARIMA model is used to identify the linear trends of gold and exchange rates. The residual values of this model are then integrated into the LSTM/GRU model for application in the hybrid framework. In addition, the forecasts of the ARIMA result and the forecasts of the LSTM/GRU model were weighted and averaged to forecast the next 3 months.

In this study, the most optimal (p, d, q) ARIMA parameters were systematically determined through the 'auto.arima' function. Thus, it provides the most effective modeling of the data. In this way, the stationarity of the time series, the autocorrelation structure, and the effect of previous error terms were tested with a large number of different values, and the process was carried out with the most appropriate one. As a result of this process, p, d, and q were determined as 3, 2, 2 respectively.

4.4 LSTM and GRU Model

In this study, a hybrid model is developed by integrating both LSTM and GRU frameworks. In the model, the LSTM and GRU layers are combined, thus providing the ability to learn both long and short-term dependencies. The hybrid model utilizes the advantages of both layer types by working with sequential data in terms of its architecture. As a result, it aims to increase the success in forecasting points.

The model was created using the Keras library in the R environment. It is a Sequential Neural Network model. The model has LSTM, GRU, Dropout, and Dense layers respectively. Each layer has different tasks to learn the complex relationships in the data and to forecast future values accurately. All of the hyperparameters in the model were determined by using the 'random search' technique and running the model iteratively many times. Then, the model went through the tuning phase, and the optimum values of all hyperparameters were tried to be obtained. Especially the difference in the number of neurons in the layers is due to this situation. The layers and hyperparameters in the model are as follows:

- LSTM Layer: It is the first layer of the model. In this layer, the long-term dependencies between the data are learned. Here, the relationships of the data over time are analyzed in depth by preserving the sequential structure of the sequential data. In this layer, 1538 neurons are used and the output is transmitted to the next layer as a sequence (return_sequences = TRUE).
- GRU Layer: The second layer in the model is the GRU layer. This layer, which is lighter and faster, is effective in learning short-term dependencies and reduces the total computational cost of the model as it requires less computational power. In this layer, 150 is used as the unit value indicating the number of neurons, and Hyperbolic Tangent (tanh) is used as the activation function. The output of this layer is a single vector (return_sequences = FALSE).
- Dropout Layer: Dropout layer was used to reduce the overfitting risk of the model. By setting rate = 0.1 in this layer, 10% of the neurons were randomly disabled during training and the generalization ability of the model was tried to be increased.
- Dense Layer: It is the layer that produces the final prediction of the model. This layer was determined to produce a single output (unit = 1) since the model is a time series prediction/regression problem.

The optimum values of the other hyperparameters in the model are loss function = Mean Squared Error (MSE), optimization algorithm = Adam (Adaptive Moment Estimation), learning rate = 0.001, training epoch = 20, and batch size = 8.

4.5 *Training and Test Data Sets*

The dataset was split into training and testing sets to facilitate the evaluation of the model's performance, as required in supervised machine learning methods. While 80% of the dataset was allocated for training, the remaining 20% was allocated for testing the model. This is done as a necessary step to evaluate the effectiveness of the model when applied to future data.

4.6 *Walk-Forward Validation*

To dynamize the forecasting process in the model, this step was taken. At this stage, the model was first trained using only the training data set, and then forecasts were made for each test data set. The model was retrained by adding the forecasted value to the training set.

4.7 *Performance Measurement and 3-Month Forecasting*

The forecasting performance of the developed hybrid model was monitored on the test data set. The prediction values found as a result of the walk-forward phase were returned to their original scales and compared with the actual values and the error metrics of the model, Mean Squared Error (MSE), Mean Absolute Error (MAE), R-squared (R^2) values were calculated. Graphical output was also obtained at this stage. As a result of this process, the outputs of the long-run forecasting performance were obtained.

In addition, it was requested to evaluate the short-term performance of the model. For this purpose, the model was subjected to one more stage to predict gold prices and exchange rates for the next 3 months. Since the success of the ARIMA model in short-term forecasting is known, a new short-term forecasting model was developed by integrating ARIMA and hybrid LSTM/GRU models. In this model, forecasts are produced by taking the weighted average of the forecasts derived from both methods. These weights are 75% and 25% for ARIMA and LSTM/GRU models, respectively. The reason for choosing these weight percentages is the superior performance of the ARIMA model in short-term forecasts compared to the LSTM/GRU model. The reliability of the model is tested by comparing the forecast results for the next 3 months with the real monthly average gold bullion sales prices and dollar and euro exchange rates for the last 3 months, which are not used in the model. The model implementation phase of this study is presented diagrammatically in Fig. 1.

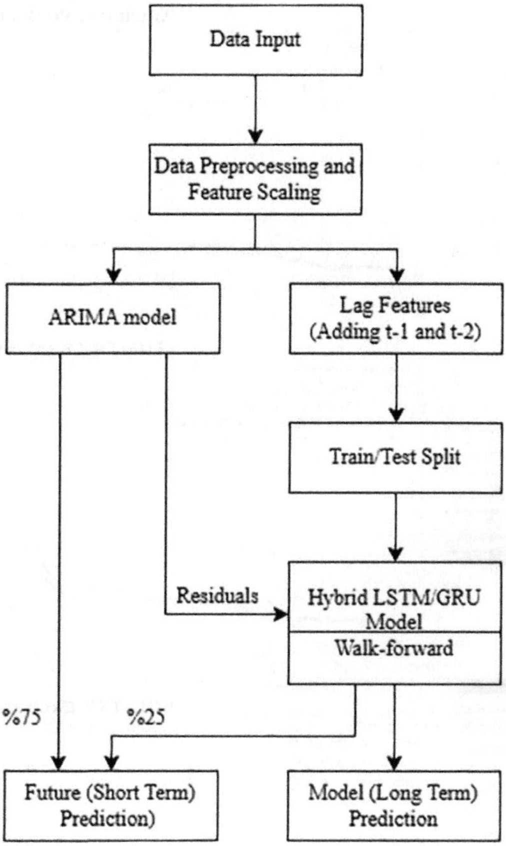

Fig. 1 Diagram of the model implementation phase

5 Results

As a result of the developed model, the performance metrics that reveal the success of the results obtained for the monthly average gold bullion sales price, Euro, and Dollar Exchange Rate at the point of accurate prediction are presented in Table 1. In addition, the comparison of both the long-term forecast values of the gold price, Euro and Dollar Exchange Rate, and the short-term forecast values, i.e. the forecast values in the next 3 months, with the actual values is shown in Fig. 2.

Table 1 Performance metrics of the hybrid model

Metric	Gold	Euro	USD
Mean squared error (MSE)	6570.073	0.955	1.101
Mean absolute error (MAE)	54.661	0.671	0.753
R-squared (R^2)	0.98	0.987	0.983

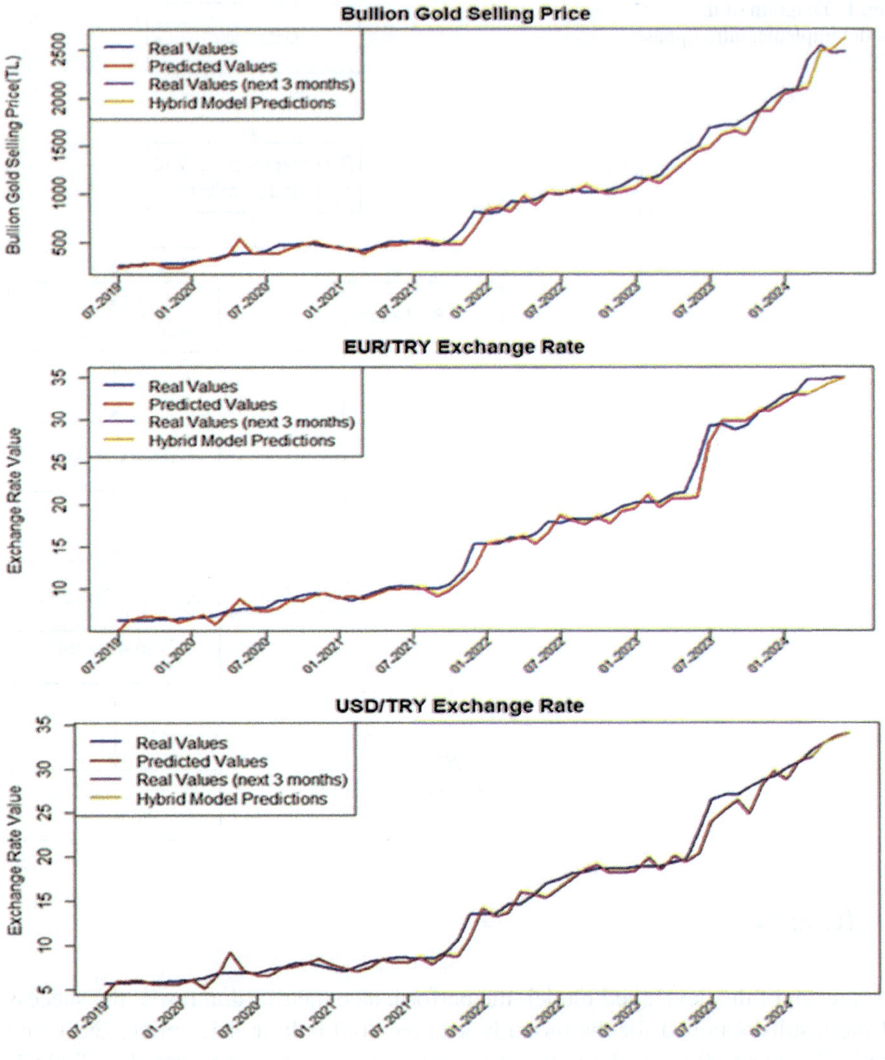

Fig. 2 Real and forecast values

5.1 Monthly Average Bullion Gold Sales Price Forecast Results

In Table 1, although the MSE value is relatively high, the R^2 value is quite high. This indicates that the model effectively captures the overall trend; however, its forecasts are subject to periodic errors. The MAE indicates that the margin of error is approximately 55 units compared to the current real price level of 2500–3000 TL, which is the average monthly gold bullion sales. This can be interpreted as a value that can be considered reasonable, especially in recent years when it is very difficult to forecast. It is observed in Fig. 2 that the actual values and the forecast values produced by the model are quite close in terms of the long term, and in terms of the future period forecast, it is observed in Fig. 2 that it shows a successful match in terms of forecasting the general trend correctly. However, in certain intervals (especially between the periods between 10–20 and 40–50 months) when the trend suddenly shifts to an upward position and maintains its position for a long time, the red forecast line shows a slight deviation from the actual values. However, even in these periods, the trend was quickly captured, but the forecast results were more cautious than the actual values. In the short-term 3-month price forecast point, the forecast was in line with the trend of the actual value, albeit cautiously, as in the long term. In summary, in general, it shows that the model accurately reflects price fluctuations, although there are some inaccuracies in terms of forecast accuracy.

5.2 USD/TRY Exchange Rate Forecast Results

When the graphical representation of the dollar exchange rate is analyzed, a significant correlation can be seen between the actual values and the forecasted values. Although the forecasts deviate from the actual figures, especially between months 45–60, the general trend is effectively captured. This period covers 2022 and beyond. Considering that the dollar exchange rate has almost doubled in this period and that it is almost impossible to forecast due to the sudden and unprecedented changes in many economic variables, it would not be wrong to say that the model still shows a very high success. In particular, the short-term forecasts for the next 3 months show a remarkable closeness to the actual values. Considering that the current price of USD/TRY exchange rate is around 34 TL, the MSE of 1.101 and the MAE of 0.753 reveals that a commendable performance has been achieved in the context of dollar pricing. The 0.983 coefficient of determination (R^2) suggests that the model performs strongly in USD forecasts. In sum, both the graphical and tabular measures suggest that the model can be successfully used to forecast the USD/TRY price in both the long and short run.

5.3 EUR/TRY Exchange Rate Forecast Results

The results of the model for the EUR/TRY exchange rate are in parallel with the results of the model for the USD/TRY exchange rate. This is a necessary and expected result. However, when analyzed in more detail, it is observed that the forecasts are more deviated compared to the dollar price forecast. In general, both short and long-term forecasts are in line with the actual values. Specifically, although the short-term forecasts for the next 3 months are slightly deviated from the actual values, they are very close to the actual values in the process. In the long-term forecasts, the general trend is captured, albeit with small temporal differences. While the MSE for the Euro was 0.955, the MAE was recorded as 0.671. Compared to the current real value of 35 TL, these error rates can be expressed as quite low. The R^2 of 0.987 indicates a very high fit, as in the other models.

When we analyze the graphs and the performance scores in the table together, it is clear that the model developed for the monthly average gold bullion sales price, dollar, and euro exchange rate forecasts gives successful and consistent results. The correlation between the actual and forecast values in the graphs is striking, and this relationship is evidenced by the measurements presented in Table 1. On the other hand, concerning the forecasts, although the model proves to be effective in determining the general trend, it exhibits slight deviations at times of uptrends. However, this deviation is rarely above the price. This can be interpreted as the model's predictions are cautious and result in a favorable outcome for the investor.

6 Conclusion

The hybrid forecasting model developed in this research provides an important solution for forecasting exchange rates and gold prices in countries like Turkey, which are characterized by frequent economic uncertainties and vulnerabilities. This hybrid method, which combines ARIMA, LSTM, and GRU models, stands out with its high accuracy in short and long-term forecasts. Thanks to its low error rates and high R^2 values, it is particularly effective in predicting complex dynamics in financial markets such as exchange rates and gold prices. One of the most striking features of the model is its ability to adapt to sudden shocks and uncertainties in the markets. This is achieved by combining the linear forecasting capabilities of ARIMA with the nonlinear learning capacity of LSTM and GRU, which allows the model to respond quickly and accurately to unpredictable financial fluctuations. In addition, the use of the walk-forward validation technique provides a dynamic structure that overcomes the limitations of traditional forecasting methods by using the most up-to-date data at each forecast stage.

While the hybrid model performs effectively overall, it has some limitations. In particular, the use of deep learning algorithms and long training times incur significant computational costs and require significant hardware resources. Furthermore, while the model is good at trend forecasting, it can sometimes lag in responding to market volatility and upward trends, suggesting that further improvements are needed.

Future research could speed up training time and reduce computational costs by more effectively balancing hyperparameter optimization with more efficient algorithms to improve the model's performance. Moreover, including more macroeconomic variables and using feature selection methods can provide a dataset that better reflects global market dynamics and speculative effects. The effectiveness and adaptability of the model under different market conditions can be tested by extending it to more stable markets globally, starting with emerging markets such as Turkey.

In conclusion, this research highlights the importance of hybrid modeling techniques in financial time series forecasting and demonstrates the advantages of combining deep learning with traditional methods. The developed model can be used as a reliable forecasting tool in markets with high economic uncertainty. Moreover, the model has the potential to be made more accurate and efficient with its adaptive structure.

References

Ashour MAH, Al-Dahhan IA (2020) Turkish lira exchange rate forecasting using time series models. IJASOS Int E J Adv Soc Sci 6(16):294–300

Bingöl S, Sadik S, Bakar RA, Mouhcine H, Albaf J, Chaabene S, Auta A (2020) Gold price prediction in times of financial and geopolitical uncertainty: a machine learning approach

Central Bank of the Republic of Turkey (n.d.) Gold prices data (TP.MK.KUL.YTL). Electronic Data Delivery System (EVDS). Retrieved 22 Oct 2024 from https://evds2.tcmb.gov.tr

Central Bank of the Republic of Turkey (n.d.) USD/TRY exchange rate data (TP.DK.USD.A.YTL). Electronic Data Delivery System (EVDS). Retrieved 8 Dec 2024 from https://evds2.tcmb.gov.tr

Central Bank of the Republic of Turkey (n.d.) EURTRY exchange rate data (TP.DK.EUR.A.YTL). Electronic Data Delivery System (EVDS). Retrieved 8 Dec 2024 from https://evds2.tcmb.gov.tr

Chandar SK, Sumathi P, Santhosh S (2016) Forecasting gold prices based on extreme learning machine. Indian J Sci Technol 9(38):1–5

Cohen G, Aiche A (2023) Forecasting gold price using machine learning methodologies. Chaos Solitons Fractals 175:114079. https://doi.org/10.1016/j.chaos.2023.114079

Cordero CA, Danao BM, Sy LA (2020) Predicting the XAU-USD foreign exchange prices using machine learning. Asian Institute of Management Research Papers

Dey R, Salem FM (2017) Gate-variants of gated recurrent unit (GRU) neural networks. In: 2017 IEEE 60th international midwest symposium on circuits and systems (MWSCAS). IEEE, pp 1597–1600. https://doi.org/10.1109/MWSCAS.2017.8053179

Elsayed N, Maida AS, Bayoumi M (2018) Deep gated recurrent and convolutional network hybrid model for univariate time series classification. arXiv preprint arXiv:1812.07683

Energy Information Administration (n.d.) Petroleum spot prices. U.S. Energy Information Administration. Retrieved 22 Oct 2024 from https://www.eia.gov/dnav/pet/pet_pri_spt_s1_m.htm

Eurostat (n.d.) Harmonised index of consumer prices (HICP). Retrieved 22 Oct 2024 from https://ec.europa.eu/eurostat/web/hicp

European Central Bank (n.d.) Key ECB interest rates. Retrieved 22 Oct 2024 from https://www.ecb.europa.eu/stats/policy_and_exchange_rates/key_ecb_interest_rates/html/index.en.html

Federal Reserve Bank of St. Louis (n.d.) Consumer price index for all urban consumers: All items in U.S. city average [CPIAUCSL]. FRED, Federal Reserve Bank of St. Louis. Retrieved 22 Oct 2024 from https://fred.stlouisfed.org/series/CPIAUCSL

Federal Reserve Bank of St. Louis (n.d.) Effective federal funds rate [FEDFUNDS]. FRED, Federal Reserve Bank of St. Louis. Retrieved 22 Oct 2024 from https://fred.stlouisfed.org/series/FEDFUNDS

Federal Reserve Bank of St. Louis (n.d.) 10-year treasury constant maturity rate [DGS10]. FRED, Federal Reserve Bank of St. Louis. Retrieved 22 Oct 2024 from https://fred.stlouisfed.org/series/DGS10

Federal Reserve Bank of St. Louis (n.d.) CBOE volatility index [VIXCLS]. FRED, Federal Reserve Bank of St. Louis. Retrieved 22 Oct 2024 from https://fred.stlouisfed.org/series/vixcls

Federal Reserve Bank of St. Louis (n.d.) Industrial production index [INDPRO]. FRED, Federal Reserve Bank of St. Louis. Retrieved 22 Oct 2024 from https://fred.stlouisfed.org/series/INDPRO

Federal Reserve Bank of St. Louis (n.d.) Interest rates, discount rate for Turkey [INTDSRTRM193N]. FRED, Federal Reserve Bank of St. Louis. Retrieved 22 Oct 2024 from https://fred.stlouisfed.org/series/INTDSRTRM193N

Ghadimpour M, Ebrahimi SB (2022) Forecasting financial time series using deep learning networks: evidence from long-short term memory and gated recurrent unit. Iran J Financ 6(4):81–94. https://doi.org/10.30699/ijf.2022.313164.1286

Giovani M, Anggriani I, Simatupang SAWD (2022) Comparison in predicting the short-term using the SARIMA, DSARIMA and TSARIMA methods. BAREKENG J Ilmu Mat Dan Ter 16(4):1487–1496

Gubler M (2012) Four essays in macroeconomic modeling of exchange rates, interest rates and commodity prices. Doctoral dissertation, University of Basel

Igboanusi MO (2024) Artificial intelligence application to EURO-USD Forex value prediction using LSTM TensorFlow. Int J Nov Res Dev 9(3):619

Investing.com (n.d.) U.S. dollar index historical data. Retrieved 22 Oct 2024 from https://tr.investing.com/indices/usdollar-historical-data

Investing.com (n.d.) BIST 100 index—historical data. Retrieved 22 Oct 2024 from https://tr.investing.com/indices/bist-100

Investing.com (n.d.) BIST 100 index—historical data (trading volume). Retrieved 22 Oct 2024 from https://tr.investing.com/indices/ise-100-historical-data

Ismail MT, Yahya A, Shabri AM (2009) Forecasting gold prices using multiple linear regression method. Am J Appl Sci 6(8):1509–1514

Jabeur SB, Mefteh-Wali S, Viviani JL (2024) Forecasting gold price with the XGBoost algorithm and SHAP interaction values. Ann Oper Res 334(1):679–699. https://doi.org/10.1007/s10479-024-05030-3

Jang H, Lee J (2019) Machine learning versus econometric jump models in predictability and domain adaptability of index options. Physica A 513:74–86. https://doi.org/10.1016/j.physa.2018.08.091

Jawadi F (2012) Introduction to time-varying modeling with macroeconomic and financial data. Macroecon Dyn 16(S2):167–175. https://doi.org/10.1017/S136510051100071X

Kilimci ZH (2022) Ensemble regression-based gold price (XAU/USD) prediction. J Emerg Comput Technol 2(1):7–12

Küçük R (2023) Forecasting foreign exchange rate with machine learning techniques. Master's thesis, Middle East Technical University

Kumar P, Priyanka P, Dhanya J, Uday KV, Dutt V (2023) Analyzing the performance of univariate and multivariate machine learning models in soil movement prediction: a comparative study. IEEE Access 11:62368–62381. https://doi.org/10.1109/ACCESS.2023.3272843

Landge U, Phokmare O, Borane N, Shelke P (2024) Gold price prediction using random forest algorithm. In: 3rd international conference on applied artificial intelligence and computing (ICAAIC). IEEE, pp 1287–1292. https://doi.org/10.1109/ICAAIC.2024.9654323

Livieris IE, Pintelas E, Pintelas P (2020) A CNN–LSTM model for gold price time-series forecasting. Neural Comput Appl 32:17351–17360. https://doi.org/10.1007/s00521-020-05179-x

Manjula KA, Karthikeyan P (2019) Gold price prediction using ensemble-based machine learning techniques. In: Proceedings of the third international conference on trends in electronics and informatics (ICOEI), pp 1360–1367

Manoj J, Suresh KK (2019) Forecast model for price of gold: multiple linear regression with principal component analysis. Thailand Statistician 17(1):125–131

Mashadihasanli T (2022) Stock market price forecasting using the ARIMA model: an application to Istanbul, Turkiye. J Econ Policy Res 9(2):439–454. https://doi.org/10.26650/jepr1056771

Mithu MA, Rahman KM, Razu RA, Riajuliislam M, Momo SI, Sattar A (2021) Gold price forecasting using regression techniques for settling economic and stock market inconsistency. In: 2021 12th international conference on computing communication and networking technologies (ICCCNT). IEEE, pp 1–4. https://doi.org/10.1109/ICCCNT49239.2021.00012

OECD (n.d.) Inflation rate: all items for Turkey. Retrieved 22 Oct 2024 from https://data.oecd.org/price/inflation-cpi.htm

OECD (n.d.) Industrial production of total industry: Euro area. Retrieved 22 Oct 2024 from https://data-explorer.oecd.org/

OECD (n.d.) Euro area unemployment rate. Retrieved 22 Oct 2024 from https://www.oecd.org/en/data/indicators/unemployment-rate.html?oecdcontrol-59006032fa-var1=EU27_2020%7CEA20&oecdcontrol-4c072e451c-var3=2024-06

Olubusola O, Mhlongo NZ, Daraojimba DO, Ajayi-Nifise AO, Falaiye T (2024) Machine learning in financial forecasting: a US review: exploring the advancements, challenges, and implications of AI-driven predictions in financial markets. World J Adv Res Rev 21(2):1969–1984

Özmen M, Şanlı S (2017) Detecting the best seasonal ARIMA forecasting model for monthly inflation rates in Turkey. Dokuz Eylül Üniversitesi İktisadi İdari Bilimler Fakültesi Dergisi 32(2):143–182. https://doi.org/10.24988/deuiibf.2017322602

Pardo R (1992) Design, testing, and optimization of trading systems, vol 2. Wiley

Pierdzioch C, Risse M, Rohloff S (2015) Forecasting gold-price fluctuations: a real-time boosting approach. Appl Econ Lett 22(1):46–50. https://doi.org/10.1080/13504851.2014.925041

Plakandaras V, Papadimitriou T, Gogas P (2015) Forecasting daily and monthly exchange rates with machine learning techniques. J Forecast 34(7):560–573. https://doi.org/10.1002/for.2346

Şahin C (2023) Mevduat faiz oranlarının ARIMA yöntemi ile tahmini: 2010–2022 dönemi Türkiye uygulaması. Sinop Üniversitesi Sosyal Bilimler Dergisi 7(1):363–384. https://doi.org/10.30561/sinopusd.1282481

Sudimanto H, Heryadi Y, Wibowo A (2021) Foreign exchange prediction using machine learning approach: a pilot study. In: 2021 4th international conference on information and communications technology (ICOIACT). IEEE, pp 239–242. https://doi.org/10.1109/ICOIACT50329.2021.9512185

Suranart K, Kiattisin S, Leelasantitham A (2014) Analysis of comparisons for forecasting gold price using neural network, radial basis function network and support vector regression. In: The 4th joint international conference on information and communication technology, electronic and electrical engineering (JICTEE). IEEE, pp 1–5. https://doi.org/10.1109/JICTEE.2014.7015717

Trading Economics (n.d.) Turkey inflation rate. Retrieved 22 Oct 2024 from https://tradingeconomics.com/turkey/inflation-cpi

Trading Economics (n.d.) Turkey—current account. Retrieved 22 Oct 2024 from https://tradingeconomics.com/turkey/current-account

Tunçsiper C (2023) Nowcasting the gold prices in Türkiye employing multilayer perceptron (MLP) networks. Int J Innov Res Multidiscip Educ 2(4). https://doi.org/10.58806/ijirme.2023.v2i4n07

Turkish Statistical Institute (n.d.) Industrial production index. Retrieved 22 Oct 2024 from https://data.tuik.gov.tr/Search/Search?text=sanayi%20%C3%BCretim%20endeksi

ul Sami I, Junejo KN (2017) Predicting future gold rates using machine learning approach. Int J Adv Comput Sci Appl 8(12). https://doi.org/10.14569/ijacsa.2017.081246

Wagh A, Shetty S, Soman A, Maste D (2022) Gold price prediction system. Int J Res Appl Sci Eng Technol 10(6):487–490. https://doi.org/10.22214/ijraset.2022.40756

Yadav DK, Soumya K, Goswami L (2024) Autoregressive integrated moving average model for time series analysis. In: International conference on optimization computing and wireless communication (ICOCWC). IEEE, pp 1–6. https://doi.org/10.1109/ICOCWC60930.2024.10470488

Yamacli DS, Yamacli S (2023) Estimation of the unemployment rate in Turkey: a comparison of the ARIMA and machine learning models including Covid-19 pandemic periods. Heliyon 9(1):e12796

Predicting the Environmental Impact of Financial Development with Machine Learning Algorithms

Burcu Kartal, Haydar Karadağ, Ahmet Şit, and Mustafa Şit

Abstract This study aimed to determine the importance levels of variables affecting carbon emissions. Various machine learning algorithms were used and their performances were compared. Energy consumption, gross domestic product (GDP), interest rate, credit volume, inflation, and uncertainty index were used as independent variables. Model performances were evaluated using MAE, MSE, RMSE, and R^2 metrics. Among the SVR, KNN, RF, ANN, XGBoost, and LightGBM algorithms, the Random Forest had the highest predictive power. The analysis revealed that the most influential variable on CO_2 emissions is energy consumption, followed by GDP, interest rate, credit volume, inflation, and uncertainty index. The results emphasize optimizing energy consumption, increasing efficiency, and switching to renewable energy to reduce carbon emissions. To ensure environmental sustainability, the study recommends increasing technology incentives, prioritizing the use of renewable energy, and policymakers to develop interest and credit policies to reduce CO_2 emissions. Thus, it states that economic growth can achieve a sustainable structure with environmentally friendly steps.

Keywords Carbon emission · Interest rate · Machine learning · Supervised learning · Ensemble learning

---

B. Kartal (✉) · H. Karadağ
The Faculty of Economics and Administrative Sciences, Recep Tayyip Erdogan University, Merkez, Rize, Türkiye
e-mail: burcu.kartal@erdogan.edu.tr

H. Karadağ
e-mail: haydar.karadag@erdogan.edu.tr

A. Şit
The Faculty of Social and Human Sciences, Malatya Turgut Ozal University, Malatya, Türkiye
e-mail: ahmet.sit@ozal.edu.tr

M. Şit
The Faculty of Tourism, Harran University, Sanliurfa, Türkiye
e-mail: msit@harran.edu.tr

M. Gün and B. Kartal (eds.), *Machine Learning in Finance*, Contributions to Finance and Accounting, https://doi.org/10.1007/978-3-031-83266-6_4

1 Introduction

Globalization brings countries closer together economically. However, the global integration process has ecological effects. Globalization has ecological impacts on countries through variables such as energy use, financial development, foreign direct investment, urbanization, and economic growth (Şeyranlıoğlu 2024: 428). Especially the increase in CO_2 emissions causes climate change and global warming (Şahin and Ünal 2023: 368). At this stage, CO_2 emissions need to be reduced due to their effects on climate change and global warming (Çil 2023: 460). However, economies need high energy consumption for large-scale industrial activities (Afşar and Yüksel 2022: 431). CO_2 emissions caused by high energy consumption have different effects on countries (Yağış and Torun 2023: 1279). Environmental problems arising from CO_2 emissions as energy consumption increases cause multidimensional damage, especially in underdeveloped and developing countries. Developed countries can combat environmental problems. However, the ability of underdeveloped and developing countries to combat environmental disasters is limited (Şahin and Ünal 2023: 368).

In the Paris Climate Agreement and IPCC (2021), it is stated that the increase in the average surface temperature of the world should be limited to 2 °C and even kept below 1.5 °C to prevent climate crisis. It was also stated in the report that CO_2 emissions that cause climate change are caused by high levels of fossil fuels (Şeyranlıoğlu 2024: 428).

Financial development, one of the variables affecting CO_2 emissions, encourages investors to invest in clean energy projects. It also reduces financial intermediation costs and provides risk diversification (Afşar and Yüksel 2022: 429). Increasing the level of financial development can attract more foreign direct capital to that country. This may increase economic growth (Şeyranlıoğlu 2024: 429). However, financial development may increase the use of fossil energy and thus environmental pollution by developing industries (Efeoğlu 2022: 2105). Financial development reduces the cost of financing by activating capital markets and facilitating access to bank loans. Thus, it can negatively affect environmental quality by increasing investment in new projects and subsequently increasing energy consumption (Çil 2023: 461; Şeyranlıoğlu 2024: 429).

From another perspective, natural resource-rich countries are less dependent on energy imports and can reduce environmental degradation by producing renewable energy. Therefore, as revenues from natural resources increase in these countries, investments in the renewable energy sector also increase (Topcu 2022: 45).

An increase in economic policy uncertainty (EPU), another variable thought to affect CO_2 emissions, affects macroeconomic indicators, financial development, and economic growth. The change in EMU also affects energy consumption and thus CO_2 emissions. An increase in EMU is expected to reduce energy consumption and CO_2 emissions by reducing production (Adedoyin and Zakari 2020: 2).

The study suggests that inflation also has positive or negative effects on CO_2 emissions. Stable inflation can boost economic growth and reduce CO_2 emissions by allocating more economic resources to environmentally friendly initiatives. However, stable inflation makes it easier for consumers to buy carbon-emitting cars, white goods, air conditioners, and durable household goods and services. Thus, it can increase energy consumption and CO_2 emissions. Inflation uncertainty, on the other hand, leads to macroeconomic instability and hinders economic growth by reducing total factor productivity and investment. Inflation uncertainty restricts the entry of firms into the market and leads to lower production and thus lower CO_2 emissions (Demirtaş 2023: 623–624).

The aim of the study is to estimate the importance levels of variables affecting CO_2 emissions using machine learning algorithms. With the estimation results, political preferences for limiting CO_2 emission at its source will be determined. The study is important as it contributes to the literature by determining the importance level of variables affecting CO_2 emission.

The study investigates whether there is a relationship between CO_2 emissions and financial development in Turkey. The introduction is followed by the Literature section, which includes studies investigating the variables affecting CO_2 emissions. This is followed by the Methodology section, which analyzes the variables affecting CO_2 emissions with a machine-learning algorithm. In the conclusion section, policy recommendations are explained in line with the results of the analysis.

2 Literature

With the process of globalization and financial liberalization, it is seen that more attention is paid to the finance-environment relationship in the economics and finance literature (Şeyranlıoğlu 2024: 429). The financial development-CO_2 relationship is researched because it supports the strengthening of the green economy and the development of clean technology (Çil 2023: 462–463). When the literature is analyzed, it is seen that studies examining the relationship between financial development and CO_2 emissions have not reached a clear conclusion (Şeyranlıoğlu 2024: 429). In the literature, the relationship between financial development and CO_2 emissions is evaluated in terms of two different results. The first one is the studies indicating that financial development and CO_2 emissions have a negative relationship. The second one is the research indicating that financial development has a very positive relationship (Yağış and Torun 2023: 1280).

Among the studies that find that financial development increases CO_2 emissions, Zhang (2011) finds that China's financial development increases CO_2 emissions with the Granger causality test in the period 1980–2009. Jiang and Ma (2019) find that financial development increases CO_2 emissions in developed countries, emerging markets, and developing countries for the period 1990–2014 with the GMM method; however, the effect of financial development on carbon emissions is insignificant in developed countries. Gök (2020) finds that financial development increases CO_2 emissions through a meta-regression analysis of 275 estimates from 72 primary studies. Sheraz et al. (2022) show that financial development significantly increases

CO_2 emissions in 64 Belt and Road Initiative countries (BRI) over the period 2003–2018 using the GMM method and Dumitrescu and Hurlin causality test. Baloch and Danish (2022) show that financial development contributes to carbon emissions in BRICS countries over the period 1995–2016 using the common correlated effect means group (CCEMG). Rani et al. (2022) found that financial development has a U-shaped relationship with CO_2 emissions and economic growth in South Asian countries (FGLS) from 1990 to 2020 using the panel data technique. Khan et al. (2022) found that financial development and economic growth increased CO_2 emissions by using the GMM method in 177 countries from 2002–2019.

Among the studies that find that financial development reduces CO_2 emissions, Zaidi et al. (2019) used the Continuously Updated Bias-Corrected (CUP-BC) and Continuously Updated Fully Modified (CUP-FM) methods for the period 1990–2016 in Asia Pacific Economic Cooperation (APEC) countries. The empirical results show that globalization and financial development have significantly reduced CO_2 emissions. Acheampong et al. (2020) show that financial market development reduces CO_2 emissions in developed and developing economies using the instrumental variable GMM method for 83 countries over the period 1980–2015. Bashir et al. (2020) find that renewable energy consumption, environmental technology, and financial development reduce CO_2 emissions in OECD economies over the period 1995–2015 using system-GMM and quantile regression approaches. Lahiani (2020) finds that an increase in financial development reduces CO_2 emissions in China for the period 1977–2013 using a unit root test with structural break and nonlinear autoregressive distributed lag model. Khan and Ozturk (2021) found that financial development reduced CO_2 emissions in 88 developing countries in the period 2000–2014 with the GMM method. Emenekwe et al. (2022) found that financial development reduces CO_2 emissions in 37 Sub-Saharan African (SSA) countries in the period 2000–2016 using the ARDL method. The Granger causality test shows that there is a bidirectional causality between financial development and CO_2 emissions.

Increased uncertainty in the economy is expected to have negative effects on macroeconomic indicators. Increased economic uncertainty will reduce the amount of energy used in production by decreasing production. Therefore, CO_2 emissions will decrease. In this context, Adedoyin and Zakari (2020), who examine the relationship between CO_2 emissions and the uncertainty index, show that EMU reduces the growth of CO_2 emissions in the short term, while the long-term use of EMU increases CO_2 emissions with the (ARDL) test for the period 1985–2017 in the United Kingdom. Atsu and Adams (2021) use ARDL (CS-ARDL) and (FMOLS) techniques for the period 1984–2017 for BRICS countries and find that policy uncertainty increases CO_2 emissions. Nakhli et al. (2022) use the Bootstrap Rolling approach for the period 1985 M1 to 2020 M12 in the USA. Empirical findings reveal that there is a bidirectional causality between CO_2 emissions and economic policy uncertainty. Liu and Zhang (2022), using a panel data model and provincial panel data for the period 2003–2017 in China, find that EMU has a negative impact on CO_2 emissions. Li et al. (2022) use a nonlinear ARDL model and asymmetric causality test for the period 2013–2021 in China. They show that four economic policy uncertainties (EPU) China's trade policy uncertainty (TPU) and monetary policy uncertainty (MPU) positively affect the carbon emission trading (CET) market price, while exchange rate policy uncertainty (ECPU) negatively affects it. Wang et al. (2022) find an inverse relationship between economic policy uncertainty (EPU) and CO_2 emission trading price (CETP) in China over the period 2013:M06-2021:M09 using a wavelet-based quantile-quantile regression approach. Benlemlih and Yavaş (2024) find that Economic Policy Uncertainty (EPU) increases firms' CO_2 emissions in 6800 firms from 23 countries. Sadiq et al. (2024) investigate the environmental Kuznets growth curve (EKGC) hypothesis for Brazil, Russia, India, and China (BRICS-1) for the period 1990–2020 using a panel cross-sectional augmented ARDL (CS-ARDL) approach. The findings show that policy uncertainty increases CO_2 emissions. Farouq et al. (2021) show that a positive shock of Financial Globalization Uncertainty (FGU) is inversely related to CO_2 emissions in nine Sub-Saharan African countries from 1980–2019. High rates of inflation cause economic uncertainty. High inflation is thought to negatively affect CO_2 emissions. Analyzing the relationship between inflation and CO_2, Grolleau and Weber (2024) apply fixed effect regressions and panel cointegration tests in 189 countries for the period 1970–2020 and find a weak but significant negative relationship between core inflation and per capita CO_2 emissions. Holding other factors constant, a 10-percentage point increase in core inflation over a 5-year period leads to a reduction in CO_2 emissions per capita of about 0.36. Rahman et al. (2024) applied the new FARDL to identify evidence of cointegration between variables. The robustness of FARDL is investigated using Bayer-Hanck joint cointegration and ARDL bounds test. There is a positive and significant effect of inflation on the carbon emission CO_2 model in Pakistan during the period 1990Q1-2016Q4. Bilal et al. (2022) The results of fully modified least squares, dynamic least squares, and robust canonical cointegration regressions in Germany over the period 1971–2016 show that alternative energy sources, government spending, and inflation are negatively related.

Tsaurai (2019), who examines the effect of loans provided by financial institutions on CO_2 emissions, finds that domestic loans provided by the financial sector increase CO_2 emissions in West African countries with the OLS method in the period 2003–2014. Acheampong (2019) finds that domestic credit provided by the financial sector to the private sector does not affect CO_2 emissions in 46 sub-Saharan African countries in the period 2000–2015 using the GMM method.

The study analyzes the variables affecting CO_2 emissions with machine learning algorithms. It contributes to the literature as it determines the importance levels of the variables affecting CO_2 emission with its findings.

3 Methodology

In the study, the data were analyzed with 6 different supervised and ensemble learning algorithms. These are Function-based SVR (Support Vector Regression) and Artificial Neural Network (ANN); Entropy-based random forest; Distance-based K-nearest neighbor (KNN); Gradient Boosting and ensemble learning algorithms XGBoost and LightGBM.

3.1 Support Vector Regression

Support Vector Regression (SVR) is the version of Support Vector Machines (SVM) proposed by Cortes and Vapnik (1995) for regression problems. In this method, data points are divided into hyperplanes. The method aims to find the hyper-plane that best fits the data and minimizes the specified error margin deviations. It is important for performance that the hyper-parameter values of the method (C, epsilon, and Kernel function) are chosen appropriately for the data. The model parameters that provide the best performance are preferred by tuning the various parameter values. SVR controls the overlearning problem and is an effective method for small datasets and nonlinear problems. It is also used for complex and high-dimensional data sets (Cortes and Vapnik 1995).

3.2 Random Forest

The random forest algorithm is one of the supervised learning and ensemble learning methods. It is used in both regression and classification problems. The method involves training multiple decision trees in parallel. The training process is done with randomly generated subsets (Bootstrap Sampling) for each tree. In classification problems, class predictions are determined as the most preferred class of each decision tree. In regression problems, the final prediction is obtained by averaging the predicted value of each decision tree. The random forest algorithm performs bagging (Bootstrap Aggregation) to minimize the model variance. It also finds the most effective variable for each node of each decision tree with a greedy search. Thus, it realizes tree branching. The fact that the method works with multiple decision trees reduces the risk of overlearning and makes it less affected by noisy data (Breiman 2001; Yang et al. 2010; Park et al. 2019).

3.3 K-Nearest Neighbor (KNN)

The k-nearest neighbor algorithm invented by Fix and Hodges (1951) was extended by T. M. Cover and P. E. Hart. K-nearest neighbor (KNN) is used in both classification and regression problems and is a popular supervised learning algorithm. KNN is a method that decides whether to classify or predict a new data point based on its k nearest neighbors (Cover and Hart 1967). The KNN algorithm first calculates the distances between the new data set and all points in the training data set. Euclidean distance method is used to calculate the distances. In the next step, it determines the k nearest neighbors selected according to the value of k, which constitutes the hyperparameter of the method. In the last step, the average value of the k neighbors is assigned as the estimated value of the new data point. The model is preferred because it is simple and flexible, makes no assumptions about the distribution, can perform multiple classifications, and is not affected by noise (Syriopoulos et al. 2023).

3.4 Artificial Neural Networks (ANN)

Artificial Neural Networks mimic biological neural networks and consist of multiple layers (input, hidden, and output layers). These layers process the data with feed-forward and back-propagation algorithms. Thanks to their multi-layered structure, neural networks can learn non-linear relationships and complex patterns. Neural networks can make powerful predictions on large data sets and complex structured data (Yegnanarayana 2006). Neural networks use a layered structure to learn the relationship between input and output (Dongare et al. 2012). They also have unique properties in terms of optimization, learning, and performance and include various learning techniques that improve the accuracy of the algorithms (Mitchell 1997).

3.5 Extreme Gradient Boosting (XGBoost)

Chen and Guestrin (2016) proposed XGBoost, which is an optimized and accelerated version of the gradient boosting algorithm. It is based on the gradient boosting framework and uses a collection of decision trees that try to correct the errors of the previous model at each step. This improves learning efficiency and performance. This method normalizes the loss function to minimize the model variance. XGBoost minimizes the loss function by generating trees with high accuracy and low complexity at each iteration. The basic steps of the XGBoost algorithm are data preparation including data preprocessing, initial tree model to determine the starting point, updating the model with gradient boosting, optimization of hyperparameter values such as max_depth, n_estimators, learning_rate, regularization to reduce the risk of overfitting, building the final model, and calculating model performance values (Chang et al.

2018). This algorithm has the advantages of speed, accuracy, and efficient results on large data sets. And it stands out for its support for incomplete data management and parallel processing (Boyko 2021). However, XGBoost has the risk of overfitting and needs accurate hyperparameter settings due to its parametric complexity (Ma et al. 2021).

3.6 Light Gradient Boosting Machine (LightGBM)

LightGBM stands for Light Gradient Boosting Machine, a gradient boosting algorithm developed by Microsoft to deliver faster and more efficient performance on large datasets. This algorithm, Gradient Boosting Decision Trees (GBDT), differs from other boosting algorithms in that it uses the Leaf-Wise Tree Growth strategy, i.e. it builds a model by expanding the deepest leaves of the tree according to the number of data (Ke et al. 2017). This approach outperforms traditional boosting algorithms such as XGBoost in terms of accuracy and speed, especially for large datasets (Cai et al. 2021). LightGBM has the advantage of faster processing times, low memory utilization, incomplete data management, and high performance on large datasets (Ju et al. 2019). However, in datasets with low data or unbalanced class distribution, the Leaf-Wise strategy can lead to overfitting. Furthermore, hyperparameter tuning can be complex and time-consuming. The LightGBM algorithm first makes a simple estimation. It then defines a function consisting of the mean square error (MSE) or the logistic loss function. It computes a gradient to calculate model errors. To speed up the optimization process, it constructs the Hesse matrix, which is the second derivative of the loss function. Identifies the data with the greatest gain when making the split decision. It creates trees thanks to the splits. After each tree is trained, it updates its predictions (Zhou et al. 2022).

4 Dataset

This study aims to estimate the variables affecting carbon emissions with machine learning algorithms. As a result of the literature review, it is seen that financial development, economic uncertainty, inflation, economic growth, and renewable energy consumption interact with carbon emissions. Carbon emissions are more problematic in developing countries than in developed countries. For this reason, analyzing the study with data from these countries will contribute to the literature. In this context, 15 countries in different regions were selected. These countries are Chile, Colombia, Czech Republic, Egypt, Hungary, Indonesia, Malaysia, Mexico, Morocco, Philippines, Poland, South Africa, Thailand, Turkey, and Vietnam. Production-based CO_2 productivity, GDP per unit of energy-related CO_2 emissions ("CO_2") for carbon emissions, Domestic credit to private sector by banks (% of GDP) ("credit") and Lending interest rate (%) ("interest") for financial development, Inflation consumer

Table 1 Descriptive statistics

Variables	N	Mean	Std. deviation	Skewness	Kurtosis
CO_2	449	4.587	1.635	0.223	0.582
Credit	449	52.739	31.503	1.159	0.833
Inflation	449	7.952	13.044	4.252	21.119
Interest	449	9.516	9.072	1.646	4.228
Uncertainty index	449	0.191	0.169	2.423	10.249
GDP	449	283.5 (Billion $)	278.0 (Billion $)	1.870	3.166
Energy	449	19.802	13.502	0.964	0.491

prices (annual %) ("inf") for inflation, Economic uncertainty ("uncertainty index") for economic uncertainty, GDP (current US$) ("GDP") for economic growth, Renewable energy consumption (% of total final energy consumption) ("energy") for energy consumption. Data were obtained from The World Bank for the period 1990–2021. Since machine learning algorithms are sensitive to missing data, unavailable country data (year and/or variable values) were excluded from the analysis. The analysis was performed with a total of 449 data. Descriptive statistics of the variables are shown in Table 1.

5 Model Implementation

5.1 Data Scaling

Variables are scaled for reasons such as improving model performance, balancing variable (attribute) scales, normalizing variable distributions, reducing the effect of extreme data, and reducing noise. For accurate prediction, the data set should be made ready for machine learning algorithms. For this reason, all variables were scaled. In the study, the max–min scaling method, which gives good results in distance-based and gradient-based algorithms, was used. This scaling reduces the data to the interval [0, 1] and improves the learning process by ensuring that the algorithm gives equal importance to each attribute.

5.2 Data Split and Validation

In supervised learning algorithms, part of the dataset is used for training, while part of the dataset is used to test the accuracy of predictions according to the proposed model. Since the prediction process is learned with the training data, this data set is heterogeneous. In this study, 80% of the dataset was selected as training data and 20% as test data. The k-fold validation method was preferred to ensure heterogeneity in the training data. In the study, k = 10 was determined.

5.3 Hyper Parameter Selections

To improve the model performance of the methods used, it is necessary to determine the appropriate hyperparameter values for the data set. The hyperparameter values for 6 different methods used in this study were obtained by grid search. Support Vector Regression hyper parameter values sigma = 1, C = 10; K-nearest neighbor hyper parameter values k = 5; random forest hyper parameter values mtry = 5; Artificial Neural Network hyper parameter values size = 15, decay = 0; XGBoost hyper parameter values booster = gbtree, eta = 0.1, max_depth = 6, gamma = 0, colsample_bytree = 0.7, min_child_weight = 1, subsample = 0.8, objective = reg:squarederror and LightGBM hyper parameter values objective = regression, metric = rmse, num_leaves = 31, learning_rate = 0.05, nrounds = 100.

5.4 Performance Measures

Mean Square Error (MSE), Mean Absolute Error (MAE), Root Mean Square Error (RMSE), and Coefficient of Determination (R^2) were used to compare the performance of the methods. MSE is calculated by averaging the squares of the differences between actual and predicted values. Since it squares the errors, it increases the impact of extreme deviations on model performance. A model with a small MSE value is a model with high predictive power. MAE is calculated by averaging the absolute values of the differences between actual values and predicted values. In MAE calculation, positive errors and negative errors do not cancel each other out. The average value of the prediction error is obtained directly. RMSE is calculated by taking the square root of MSE. It is obtained by taking the square root of the mean of the error squares between the actual values and the predicted values. Although it is as sensitive to larger errors as MSE, it is more comprehensible than MSE because it takes the square root. R^2 is the percentage of the total variance explained by the model for the independent variables. It measures how well the model explains the data. A high value of R^2 gives the model high explanatory power.

6 Findings

This study utilizes six different machine learning methods to estimate the impact of variables affecting carbon emissions. The results comparing the methods evaluated according to various performance criteria are given in Table 2.

According to Table 2, Random Forest performed the best in carbon emission estimation with the lowest MAE (0.0374) and RMSE (0.0595) values. It is also the most successful model in explaining the carbon emission of independent variables with the highest R^2 value (0.9034). XGBoost performed well with low error rates and high R^2 (0.8666), but slightly lower than the Random Forest. SVR and KNN have similar error rates and explanatory power, following Random Forest and XGBoost. ANN and LightGBM are the poorest-performing models with high error rates and low R^2 values.

The most appropriate model for the created data set was determined as the Random Forest model. It was decided to evaluate the results specific to this method. In the Random Forest model, carbon emission estimates deviate by 0.0374 units on average. This value indicates that the model's predictions have a high accuracy rate. The low MAE value indicates that the model does not make a systematic error and has a strong generalization ability. The Random Forest model shows a relative error of 28.60% in carbon emission estimates. This result indicates that the model is consistent in its predictions compared to actual values and is able to effectively model complex relationships between variables. In terms of relative error rate, Random Forest outperforms the other methods. Moreover, the RMSE calculated at 0.0595 indicates that the model's error level in carbon emission estimates is low. A low RMSE value indicates that the model predicts complex relationships with high accuracy and that the model avoids overlearning. The R^2 value was calculated as 0.9034. This means that R^2 explains 90.34% of the variance of the independent variables on carbon emissions. This high R^2 value indicates that the model provides a strong relationship between the independent variables and carbon emissions and a high level of explanatory power. Thus, Random Forest was able to successfully capture the non-linear and complex relationships between the independent variables and carbon emissions. In addition, it provided a high consistency between the training and test datasets and avoided overfitting.

Table 2 Comparison of model performances

Model	MAE	MRE	RMSE	R^2
SVR	0.0438	0.2404	0.0683	0.8482
Random Forest	0.0374	0.2860	0.0595	0.9034
KNN	0.0433	0.2686	0.0689	0.8481
ANN	0.0497	0.2217	0.0754	0.8134
XGBoost	0.0437	0.2656	0.0662	0.8666
LightGBM	0.0529	0.3539	0.0778	0.8099

Table 3 Importance levels of independent variables

Variables	Importance levels (%)
Energy	28.39
GDP	22.03
Interest	19.06
Credit	13.46
Inflation	11.57
Uncertainty index	5.50

The importance of the independent variables used in carbon emission estimation was assessed with the Random Forest model. This analysis shows that energy consumption is the strongest determinant of carbon emissions. The relative importance levels of the variables are given in Table 3.

Table 3 shows that energy consumption is the most critical variable in estimating carbon emissions. This finding emphasizes the direct impact of energy consumption on carbon emissions. GDP (22.03%) shows a strong relationship between economic growth and carbon emissions. Interest rate, credit volume, and inf reveal the environmental impact of economic and financial systems through their effects on carbon emissions. Uncertainty index has a relatively weak effect on carbon emissions.

7 Conclusion and Recommendation

This study, which estimates the impact of variables affecting carbon emissions, uses various machine learning algorithms and evaluates performance on different metrics. Carbon emission is the dependent variable. The independent variables are energy consumption, gross domestic product (GDP), interest rate, credit volume, inflation, and uncertainty index. The data set was preprocessed using min–max normalization and hyper-parameter optimization was applied. According to MAE, MSE, RMSE, and R^2 performance criteria, SVR, KNN, Random Forest, ANN, XGBoost, and LightGBM methods were evaluated, and Random Forest was found to have the best predictive power. The results of the random forest algorithm show that energy, GDP, interest, credit, inf, and uncertainty index variables contribute the most to the estimation of the CO_2 emission variable. The fact that the most influential variable is the energy variable indicates that more energy use results in more carbon dioxide use (Zaman and Mitwali 2017; Alshehry and Belloumi 2015). Countries aiming for economic growth must increase their production capacity. However, it is inevitable to use more energy for this. More energy use will increase CO_2 emissions into the atmosphere (Aktar et al. 2021; Feng et al. 2011). In fact, it is stated that energy use and CO_2 emissions occur mostly in the industrial and transportation sectors (Paul and Bhattacharya 2004). This shows how important the growth rates of countries are for the country's economy. It is necessary to generate income for economic growth.

According to the level of development of countries, the biggest sources of income are production and service revenues. This income composition differs according to the level of development of countries. At the micro level, it is observed that segments with higher incomes lead to more energy consumption and CO_2 emissions (Feng et al. 2011). This can be explained both by the fact that they produce more to generate more income and that they switch to luxury consumption and increase energy consumption as their income level increases.

After GDP, interest is the variable that affects the model the most. The cost of funding in countries varies depending on the policy interest rate. A decrease in the cost of funding has a positive effect on the profitability of enterprises. Individual or corporate loans lead to more carbon emissions (Kim et al. 2020). However, if loans from fund-supplying institutions are allocated to consumption instead of production, demand inflation rises in the country's markets (Goodhart et al. 2023).

Increasing bank lending in a country increases carbon emissions. In particular, a decrease in credit costs increases the volume of market bank loans. In other words, there is an inverse relationship between credit costs and CO_2 emissions (Zhu and Zhao 2022; Kacperczyk and Peydró 2022). When firms' use of more bank loans is channeled into investment, their production will increase. More production requires energy. This will increase CO_2 emissions.

In periods of high inflation and uncertainty, investors may be discouraged from investing (Muritala 2011) and may prefer to hold their funds in more liquid assets (Duchin et al. 2010). This may lead to a contraction in the markets. In addition, during periods of inflation and uncertainty, the supply chain of enterprises narrows due to decreases in demand and increases in costs. Enterprises reduce their production capacity. Thus, CO_2 emissions decrease (Ullah et al. 2020; Adams et al. 2020).

The results show that energy consumption, economic growth, interest rates, and bank lending have significant effects on CO_2 emissions. Technology incentives to reduce energy consumption and increase efficiency will reduce CO_2 emissions. It is also important to align economic growth processes with sustainability.

The increase in energy costs, especially in recent years, reveals the importance of renewable energy supply. This situation encourages countries to produce renewable energy. It can be said that Energy will be more important in the growth of countries in the coming years. More energy use will lead to more CO_2 emissions (Tucker 1995; Marjanović et al. 2016; Ramanathan 2006). Revenues from the production of high-value-added products will be sustainable for the national economy. However, against the negative impact of CO_2 emissions on the atmosphere and environmental pollution, environmentally friendly steps should be taken for the country's economy. Especially in energy production, energy production with renewable resources should be emphasized instead of resources that increase CO_2 emissions. In addition, funding costs should be reduced by policymakers in the country to reduce CO_2 emissions. Countries should develop inflation, interest rates, and credit policies that will support environmental sustainability in the financial development process.

References

Acheampong AO (2019) Modeling for Insight: does financial development improve environmental quality? Energy Econ 83:156–179. https://doi.org/10.1016/j.eneco.2019.06.025

Acheampong AO, Amponsah M, Boateng E (2020) Does financial development mitigate carbon emissions? evidence from heterogeneous financial economies. Energy Econ 88:104768. https://doi.org/10.1016/j.eneco.2020.104768

Adams S, Adedoyin F, Olaniran E, Bekun FV (2020) Energy consumption, economic policy uncertainty, and carbon emissions; causality evidence from resource-rich economies. Econ Anal Policy 68:179–190

Adedoyin FF, Zakari A (2020) Energy consumption, economic expansion, and CO_2 emission in the UK: the role of economic policy uncertainty. Sci Total Environ 738:140014. https://doi.org/10.1016/j.scitotenv.2020.140014

Afşar M, Yüksel GÖ (2022) Küreselleşme, Finansal Gelişme ve Karbon Emisyonları İlişkisi: Türkiye Üzerine Asimetrik Kanıtlar. Sakarya İktisat Dergisi 11(4):428–449

Aktar MA, Alam M, Al-Amin AQ (2021) Global economic crisis, energy use, CO_2 emissions, and policy roadmap amid COVID-19. Sustain Prod Consum 26:770–781

Alshehry AS, Belloumi M (2015) Energy consumption, carbon dioxide emissions, and economic growth: the case of Saudi Arabia. Renew Sustain Energy Rev 41:237–247. https://doi.org/10.1016/j.rser.2014.08.004

Atsu F, Adams S (2021) Energy consumption, finance, and climate change: does policy uncertainty matter? Econ Anal Policy 70:490–501. https://doi.org/10.1016/j.eap.2021.03.013

Baloch MA, Danish, (2022) CO_2 emissions in BRICS countries: what role can environmental regulation and financial development play? Climatic Change 172(1):9. https://doi.org/10.1007/s10584-022-03362-7

Bashir MF, Benjiang MA, Shahbaz M, Jiao Z (2020) The nexus between environmental tax and carbon emissions with the roles of environmental technology and financial development. PLoS One 15(11):e0242412. https://doi.org/10.1371/journal.pone.0242412

Benlemlih M, Yavaş ÇV (2024) Economic policy uncertainty and climate change: evidence from CO_2 emission. J Bus Ethics 191:415–441. https://doi.org/10.1007/s10551-023-05389-x

Bilal IK, Tan D, Azam W, Hassan ST (2022) Alternate energy sources and environmental quality: the impact of inflation dynamics. Gondwana Res 106:51–63. https://doi.org/10.1016/j.gr.2021.12.011

Boyko N (2021) Research into machine learning algorithms for the construction of mathematical models of multimodal data classification problems. Comput Probl Electr Eng 11(2):1–11

Breiman L (2001) Random forests. Mach Learn 45(1):5–32

Cai J, Li X, Tan Z, Peng S (2021) An assembly-level neutronic calculation method based on the LightGBM algorithm. Ann Nucl Energy 150:107871

Chang YC, Chang KH, Wu GJ (2018) Application of eXtreme gradient boosting trees in the construction of credit risk assessment models for financial institutions. Appl Soft Comput 73:914–920

Chen T, Guestrin C (2016) Xgboost: a scalable tree boosting system. In: Proceedings of the 22nd ACM SIGKDD International conference on knowledge discovery and data mining, KDD'16. ACM, New York, pp 785–794

Cortes C, Vapnik V (1995) Support-vector networks. Mach Learn 20(3):273–297

Cover TM, Hart PE (1967) Nearest neighbor pattern classification. IEEE Trans Inf Theory IT 13(1):21–27

Çil D (2023) Ekonomi ve çevre etkileşimi: Türkiye'de finansal gelişim ve ulaşım kaynaklı CO_2 ilişkisi üzerine Fourier ADL analizi. *Sosyal Mucit Academic Review* 4(4):456–476. https://doi.org/10.54733/smar.1402556

Demirtaş C (2023) Does inflation instability cause environmental pollution in Türkiye? evidence from the fourierwavelet causality test. J Res Econ Polit Financ 8(4):622–642. https://doi.org/10.30784/epfad.1353763

Dongare AD, Kharde RR, Kachare AD (2012) Introduction to artificial neural network. Int J Eng Innov Technol (IJEIT) 2(1):189–194

Duchin R, Ozbas O, Sensoy BA (2010) Costly external finance, corporate investment, and the subprime mortgage credit crisis. J Financ Econ 97(3):418–435

Efeoğlu R (2022) Çevresel Kuznets Eğrisi Çerçevesinde Sanayileşme, Yenilenebilir Enerji, Enerji Tüketimi ve Finansal Gelişmenin CO_2 Salınımı Üzerindeki Etkisi. Alanya Akademik Bakış 6(2):2103–2115

Emenekwe CC, Onyeneke RU, Nwajiuba CU (2022) Financial development and carbon emissions in Sub-Saharan Africa. Environ Sci Pollut Res 29:19624–19641. https://doi.org/10.1007/s11356-021-17161-7

Farouq IS, Sambo NU, Ahmad AU, Jakada AH, Danmaraya IA (2021) Does financial globalization uncertainty affect CO_2 emissions? empirical evidence from some selected SSA countries. Quant Financ Econ 5(2):247–263. https://doi.org/10.3934/QFE.2021011

Feng ZH, Zou LL, Wei YM (2011) The impact of household consumption on energy use and CO_2 emissions in China. Energy 36(1):656–670

Fix E, Hodges JL (1951) Discriminatory analysis, nonparametric discrimination, consistency properties. Project 21-49-004. Report No.4 USAF School of Aviation Medicine Randolph Field, Texas, USA, pp 1–21

Goodhart CA, Tsomocos DP, Wang X (2023) Bank credit, inflation, and default risks over an infinite horizon. J Financ Stab 67:101131

Grolleau G, Weber C (2024) The effect of inflation on CO_2 emissions: an analysis over the period 1970–2020. Ecol Econ 217:108029. https://doi.org/10.1016/j.ecolecon.2023.108029

Gök A (2020) The role of financial development on carbon emissions: a meta regression analysis. Environ Sci Pollut Res 27:11618–11636. https://doi.org/10.1007/s11356-020-07641-7

IPCC—Intergovernmental Panel Climate Change (2021) IPCC Press Release, 2021/17/PR

Jiang C, Ma X (2019) The impact of financial development on carbon emissions: a global perspective. Sustainability 11(19):5241. https://doi.org/10.3390/su11195241

Ju Y, Sun G, Chen Q, Zhang M, Zhu H, Rehman MU (2019) A model combining convolutional neural network and LightGBM algorithm for ultra-short-term wind power forecasting. IEEE Access 7:28309–28318

Kacperczyk MT, Peydro JL (2022) Carbon emissions and the bank-lending channel. European Corporate Governance Institute—Finance Working Paper No. 991/2024. https://doi.org/10.2139/ssrn.3915486

Ke G, Meng Q, Finley T, Wang T, Chen W, Ma W, Liu TY (2017) LightGBM: a highly efficient gradient boosting decision tree. Adv Neural Inf Process Syst 30:3146–3154

Khan H, Weili L, Khan I (2022) Institutional quality, financial development and the influence of environmental factors on carbon emissions: evidence from a global perspective. Environ Sci Pollut Res 29:13356–13368. https://doi.org/10.1007/s11356-021-16626-z

Khan M, Ozturk I (2021) Examining the direct and indirect effects of financial development on CO_2 emissions for 88 developing countries. J Environ Manage 293:112812. https://doi.org/10.1016/j.jenvman.2021.112812

Kim DH, Wu YC, Lin SC (2020) Carbon dioxide emissions and the finance curse. Energy Economics 88:104788

Lahiani A (2020) Is financial development good for the environment? an asymmetric analysis with CO_2 emissions in China. Environ Sci Pollut Res 27:7901–7909. https://doi.org/10.1007/s11356-019-07467-y

Li X, Li Z, Su CW, Umar M, Shao X (2022) Exploring the asymmetric impact of economic policy uncertainty on China's carbon emissions trading market price: do different types of uncertainty matter? Technol Forecast Soc Chang 178:121601. https://doi.org/10.1016/j.techfore.2022.121601

Liu Y, Zhang Z (2022) How does economic policy uncertainty affect CO_2 emissions? a regional analysis in China. Environ Sci Pollut Res 29:4276–4290. https://doi.org/10.1007/s11356-021-15936-6

Ma M, Zhao G, He B, Li Q, Dong H, Wang S, Wang Z (2021) XGBoost-based method for flash flood risk assessment. J Hydrol 598:126382

Marjanović V, Milovančević M, Mladenović I (2016) Prediction of GDP growth rate based on carbon dioxide (CO_2) emissions. J CO_2 Util 16:212–217

Mitchell TM (1997) Machine learning. Carnegie Mellon University

Muritala T (2011) Investment, inflation and economic growth: empirical evidence from Nigeria. Res J Financ Account 2(5):68–76

Nakhli MS, Shahbaz M, Jebli MB, Wang S (2022) Nexus between economic policy uncertainty, renewable & non-renewable energy and carbon emissions: contextual evidence in carbon neutrality dream of USA. Renew Energy 185:75–85. https://doi.org/10.1016/j.renene.2021.12.046

Park S, Hamm SY, Kim J (2019) Performance evaluation of the GIS-based datamining techniques decision tree, random forest, and rotation forest for landslide susceptibility modeling. Sustainability 11(20):5659

Paul S, Bhattacharya RN (2004) CO_2 emission from energy use in India: a decomposition analysis. Energy Policy 32(5):585–593

Rahman SU, Faisal F, Sami F, Ali A, Chander R, Amin MY (2024) Investigating the nexus between inflation, financial development, and carbon emission: empirical evidence from FARDL and frequency domain approach. J Knowl Econ 15:292–318. https://doi.org/10.1007/s13132-022-01076-w

Ramanathan R (2006) A multi-factor efficiency perspective to the relationships among world GDP, energy consumption and carbon dioxide emissions. Technol Forecast Soc Chang 73(5):483–494

Rani T, Amjad MA, Asghar N, Rehman HU (2022) Revisiting the environmental impact of financial development on economic growth and carbon emissions: evidence from South Asian economies. Clean Techn Environ Policy 24:2957–2965. https://doi.org/10.1007/s10098-022-02360-8

Sadiq M, Hassan ST, Khan I, Rahman MM (2024) Policy uncertainty, renewable energy, corruption and CO_2 emissions nexus in BRICS-1 countries: a panel CS-ARDL approach. Environ Dev Sustain 26:21595–21621. https://doi.org/10.1007/s10668-023-03546-w

Sheraz M, Deyi X, Mumtaz MZ, Ullah A (2022) Exploring the dynamic relationship between financial development, renewable energy, and carbon emissions: a new evidence from belt and road countries. Environ Sci Pollut Res 29:14930–14947. https://doi.org/10.1007/s11356-021-16641-0

Syriopoulos PK, Kalampalikis NG, Kotsiantis SB, Vrahatis MN (2023) KNN classification: a review. Ann Math Artif Intell, 1–33

Şahin C, Ünal S (2023) Karbon Salınımının, Ekonomik Büyüme ve Finansal Gelişme ile İlişkisi: Türkiye Örneği. Akademik Araştırmalar ve Çalışmalar Dergisi 15(29):367–376. https://doi.org/10.20990/kilisiibfakademik.1317807

Şeyranlıoğlu O (2024) Türkiye'de Finansal Gelişmenin Karbon (CO2) Emisyonlarına Etkisi: Çevresel Kuznets Eğrisi Hipotezi Çerçevesinde Sektörel Bir Bakış. Sosyoekonomi 32(61):427–454

Topcu BA (2022) Türkiye'de Sürdürülebilir Çevre: Finansal Gelişme, Yenilenebilir ve Yenilenemeyen Enerji Tüketimi ve Doğal Kaynak Gelirlerinin Rolü. *Akademik Sosyal Araştırmalar Dergisi* 10(129):43–60. https://doi.org/10.29228/ASOS.61995

Tsaurai K (2019) The impact of financial development on carbon emissions in Africa. Int J Energy Econ Policy 9(3):144–153

Tucker M (1995) Carbon dioxide emissions and global GDP. Ecol Econ 15(3):215–223

Ullah S, Apergis N, Usman A, Chishti MZ (2020) Asymmetric effects of inflation instability and GDP growth volatility on environmental quality in Pakistan. Environ Sci Pollut Res 27:31892–31904

Wang KH, Liu L, Zhong Y, Lobonţ OR (2022) Economic policy uncertainty and carbon emission trading market: a China's perspective. Energy Econ 115:106342. https://doi.org/10.1016/j.eneco.2022.106342

Yağış O, Torun M (2023) Enerji Tüketimi, Çevre Kalitesi ve Finansal Gelişme Arasındaki İlişkinin İncelenmesi: Panel Eşanlı Denklemler Sistemi Analizi. Yönetim Bilimleri Dergisi 21(50):1277–1294. https://doi.org/10.35408/comuybd.1315304

Yang P, Yang YH, Zhou BB, Zomaya AY (2010) A review of ensemble methods in bioinformatics. Curr Bioinform 5(4):296–308

Yegnanarayana B (2006) Artificial neural networks. Prentice-Hall of India

Zaidi SAH, Zafar MW, Shahbaz M, Hou F (2019) Dynamic linkages between globalization, financial development and carbon emissions: evidence from Asia Pacific economic cooperation countries. J Clean Prod 228:533–543. https://doi.org/10.1016/j.jclepro.2019.04.210

Zaman K, Mitwali MA (2017) Energy consumption, carbon dioxide emissions and economic development: Evaluating alternative and plausible environmental hypothesis for sustainable growth. Renew Sustain Energy Rev 74:1119–1130. https://doi.org/10.1016/j.rser.2017.02.072

Zhang YJ (2011) The impact of financial development on carbon emissions: an empirical analysis in China. Energy Policy 39(4):2197–2203. https://doi.org/10.1016/j.enpol.2011.02.026

Zhou Y, Wang W, Wang K, Song J (2022) Application of LightGBM algorithm in the initial design of a library in the cold area of China based on comprehensive performance. Buildings 12(9):1309

Zhu B, Zhao Y (2022) Carbon risk and the cost of bank loans: evidence from China. Technol Forecast Soc Chang 180:121741

A Comparative Analysis of Artificial Neural Networks and Time Series Models in Exchange Rate Forecasting

Emre Ürkmez

Abstract This study aims to compare the performances in intra-period forecast values of the exchange rate, which is one of the classic linear time series models, ARIMA model, and one of machine learning methods, artificial neural network models. This study uses monthly data for the USD/TRY exchange rates for Turkey over the period from January 2010 to October 2024. Exchange rates are sensitive to unexpected events and political uncertainties in the economy. It is, therefore, highly considerable that the estimation of the future values of exchange rates is important to central banks and companies concerning risk management. ARIMA is a linear univariate time series model that makes in-sample and/or out-of-sample value estimates based on the AR and MA components in the relevant time series data. ARIMA models are models based on a linear component of past data that makes estimates and can yield statistically reliable results. However, all these models are based on the assumption of data having a linear structure and do not give successful results in estimates against nonlinear series. In the literature, it was indicated that the ANN model is more successful than ARIMA models in capturing and predicting the nonlinear structure of the data structure. Since the USD/TRY exchange rate might have a nonlinear structure due to this fact, the use of the ANN model, methods were preferred in this study. According to the empirical findings in the study, the intra-period forecast values obtained with the ANN model on the USD/TRY exchange rate were determined to be more successful than the ARIMA model according to the forecast performance criteria.

Keywords ANN · ARMA · Exchange rate · Finance · Forecasting · Machine learning

E. Ürkmez (✉)
Department of Economics, Recep Tayyip Erdogan University, Rize, Türkiye
e-mail: emre.urkmez@erdogan.edu.tr

M. Gün and B. Kartal (eds.), *Machine Learning in Finance*, Contributions to Finance and Accounting, https://doi.org/10.1007/978-3-031-83266-6_5

1 Introduction

Since the 1970s, starting from developed countries, a floating exchange rate regime has been adopted and later adopted by developing countries, leading to a rapid growth in world trade volumes, and foreign exchange markets have undergone significant expansion. As the main determinants of international trade, exchange rates play a critical role in shaping the import–export sectors in particular (Kamruzzaman and Sarker 2004). For example, when the US Dollar gains value against the Turkish Lira, firms located in the US can procure Turkish goods and services at a more affordable cost. This increases Turkey's exports to the US and leads to a decrease in its imports. On the other hand, when the US Dollar loses value against the Turkish Lira, goods and services imported from the US become more affordable for firms and households located in Turkey. This increases Turkey's imports to the US and other countries. As a result, an increase or decrease in exchange rates may lead countries to seek alternative trading partners, and this may reflect the profound effects of permanent changes in exchange rates on trade flows and market decisions (McKenzie 2002; Hau and Rey 2005).

International trade relations are inherently dependent on the exchange rate parity of countries' local currencies against other countries' currencies due to the costs associated with exports and imports (Faini and Melo 1990; Lane and Ferretti 2002). Although international trade agreements vary from short-term agreements to long-term commitments, exchange rates fluctuate continuously at all times during the days and hours when foreign exchange markets are open. Therefore, accurate estimation of exchange rates is important for firms, investors, and financial institutions engaged in international trade. Accurate exchange rate estimations contribute to effective risk management by enabling each market participant to make price optimizations (Levich 1985; Papaioannou 2006).

Despite the volatility and uncertainty of exchange rate price movements, exchange rate forecasting has led to the development of a variety of forecasting methods, from classical univariate time series forecasting models to machine learning methods and artificial neural networks. These methods reflect the efforts of researchers and international firms to forecast exchange rate price movements and are important to the critical role of market makers in maintaining international trade and financial stability.

There are some difficulties in making reliable estimates of financial time series. Contrary to the assumptions of traditional econometric methods, financial time series are not normally distributed (Brooks 2019). The distribution of financial time series generally deviates from the normal distribution due to the thickness and extreme sharpness of the tails. The thickness of the tails corresponds to the points where there are extreme, large price movements in the financial time series. In contrast, extreme sharpness indicates that extreme values in the financial time series are expected with a higher probability. Extreme values in financial time series correspond to periods of high volatility (Brooks 1998). Since financial time series have these two properties, they exhibit leptokurtic distribution (Carol 2009; Yavuz 2014). Therefore, the distribution of financial time series tends to exhibit distribution properties that are more pointed on average and fatter in the tails than the normal distribution. Due to these properties, working with univariate time series models may not be suitable for making accurate and reliable estimates (Clements and Hendry 2000).

The autoregressive integrated moving average (ARIMA) model, which forms the basis of traditional time series analysis, is the most widely used linear forecasting method. The stationarity of ARIMA models depends on the stationarity of the autoregressive part of the model. It is assumed based the assumption on that the time series under study has a linear structure. These assumptions model the series under study through the autoregressive (AR) and moving average (MA) components of the ARMA model and obtain effective estimates (Zhang 2003). If the mean and variance of the time series under study change over time and are not linear, the applicability and effectiveness of the ARMA model are significantly weakened. In the studies conducted in the literature, exchange rate data exhibit a nonlinear and chaotic structure. Deviations from these basic assumptions make it inappropriate to work with the ARMA models (Moosa 2000; Spyros and Nikolas 2016). This situation requires alternative forecasting techniques that can better accommodate the non-linear and chaotic features of volatile financial data sets, especially exchange rates.

Considering this disadvantageous situation of the ARMA models, in recent years' financial time series have been estimated using artificial neural networks (ANN) models with machine learning methods. The ANN models, known for modeling nonlinear structures and adapting to high-frequency series, offer an alternative to the ARMA models, which are traditional forecasting methods. This study aims to compare the performances in intra-period forecast values of the exchange rate, which is one of the traditional linear time series models, the ARMA model, and one of the machine learning methods, artificial neural network (ANN) models. For this purpose, the USD/TRY exchange rate is estimated separately for both models and the forecasting performance criteria are taken into consideration and it is analyzed which model has better forecasting performance.

The remaining part of this study is organized in four parts: the second section summarizes the literature studies and findings about the data on the exchange rate; the third section presents the data of the USD/TRY exchange rate used in the analysis and then explains the traditional time series model of ARMA and the machine learning method of ANN. Second, in the chapter, the forecasting performance measures are briefly summarized to evaluate the effectiveness of two different alternative models used within the study. The findings obtained from both the ARMA and ANN models are presented and interpreted in the fourth section. Lastly, the findings of the study are evaluated and discussed in the fifth section. The limitations are also discussed, and further studies are recommended.

2 Literature Review

The studies in the literature have examined that it is seen that machine learning methods are used less in exchange rate forecasting than traditional time series forecasting models. The ANN models have significant advantages over traditional forecasting methods. ANN models are successful in solving and predicting nonlinear structures where ARIMA models are inadequate. ANN models do not make any assumptions about the statistical distribution of the data. This makes ANN models flexible and powerful models in the presence of missing data and outliers. In addition, ANN models differ from traditional methods by continuously learning from data and improving predictive performance.

This study only includes studies where exchange rates are estimated with the ANN models. Franses and Homelen (1998) aimed to model and forecast four exchange rate returns with respect to the Dutch guilder using ANN models. Simulations showed that neglecting the GARCH effect did not produce successful ANN results and that ANNs would use nonlinear structures other than GARCH to improve forecasts. They found that the forecasting performance of ANNs was not strong in the sample data. These results indicated that the nonlinear structure in exchange rates was most likely due to GARCH and suggested that ANNs could be used to diagnose average nonlinearity. Kamruzzaman and Sarker (2004) examined neural network models based on three learning algorithms: Standard Backpropagation (SBP), Scaled Conjugate Gradient (SCG), and Bayesian Regularized Backpropagation (BPR) for forecasting exchange rates against the Australian dollar. Using five technical indicators, they found that the SCG model outperformed two measures and performed comparably to BPR in the remaining three, demonstrating that effective forecasts can be achieved with basic technical indicators. According to Panda and Narasimhan (2007), the ANN model follows the step forward in predicting the weekly INR/US Dollar exchange rate. In this study, the precision of a neural network was compared against traditional linear autoregressive moving average models. The neural network outperformed other models for both in-sample and out-of-sample predictions based on six different evaluation criteria. These results provide evidence against the efficient market hypothesis and suggest that hidden information in exchange rates can be used in future predictions. Kadılar et al. (2009) suggested using the ANN for the high volatility Turkish Lira/US Dollar exchange rate series in their study and the results showed that the ANN provides the highest forecast accuracy compared to seasonal ARIMA and ARCH time series models. They also provided detailed suggestions on the use of the ANN in Turkish exchange rate forecasting. Pradhan and Kumar (2010) apply the ANN model for the forecast of the foreign exchange for India. The study used on Euro, US Dollar, Japanese yen, and British pound for daily and monthly data. The measures of forecast error are considered while assessing the performance of the forecast such as RMSE, MAE, MAD, and MAPE. Their findings suggest that ANN is a viable option when conducting forecasts of the exchange rate. By using a monthly data set between January 2000 and September 2014, Aydın and Cavdar (2015) estimated the USD/TRY exchange rate, gold price, and the BIST 100 index, some of the key macroeconomic variables. ANN and VAR models were used in this study. The estimated results of the ANN model indicated that the financial crisis in the BIST 100 index and the USD/TRY exchange rate may start in October 2017 excluding gold price. The findings concluded that the ANN model gave better prediction values than the VAR model, even though there were financial crises in the analyzed period. Kristjanpoller and Hernandez (2017) analyzed the series of silver, gold, and copper spot prices between September 7, 1999, and May 20, 2014, using daily data. In the study, they estimated the volatility of the returns of three precious metals using a hybrid neural network and GARCH-type models while analyzing the series. The findings concluded that only the hybrid neural network model successfully estimated the volatility of the returns of the spot prices of three precious metals. Sun et al. (2020) analyzed the time series data of four major currencies, that is, the US Dollar against the Japanese Yen, Chinese Yuan, British Pound, and Euro, using daily data from January 3, 2011 to December 29, 2017. The findings showed that the Bagging-based LSTM (LSTM-B) ensemble deep learning approach made better predictions than alternative models in forecasting exchange rates. Chen et al. (2021) investigated if the economic and technological determinants might serve to generate accurate predictions of Bitcoin exchange rates. They applied ANN and Random Forest methods to judge the importance of those determinants and reviewed Bitcoin exchange rate predictions by using other models like ANFIS, ARIMA, SVR, and LSTM. From their results, it was perceived that economic and technological determinants effectively affected or influenced Bitcoin exchange rate estimations in various periods. Charef (2024) has suggested the use of the ANN-GARCH model to estimate daily exchange rate return volatility. The results showed that the ANN-GARCH model could be an alternative to a conventional linear autoregressive model.

In the literature, the works involving machine learning methods in the prediction of exchange rates against the Turkish Lira are minimal. In the present study, using the monthly data from 2010 through 2024, an in-sample forecast of the exchange rate of the USD/TRY for Turkey was realized. In this study, two different models, which are the ANN model representing machine learning methods and the ARIMA model, a traditional times series model, have been used. In this regard, the current USD/TRY exchange rate data was used to compare the intra-period prediction performances estimated by two different alternative forecast models to reinforce the related studies.

3 The Data and Methodology

3.1 *The Data*

In this study, intra-period estimates of the USD/TRY exchange rate (ER) were made using a monthly dataset for Turkey over the period from January 2010 to October 2024. The USD Dollar Turkish Lira exchange rate data set used in the study was collected from Trading Economics.

3.2 The Methodology

In this section of the study, firstly the ARIMA model, which is one of the linear time series models, is explained. Then, an alternative forecasting method, machine learning, and the ANN model are given. Finally, for the comparison, the achievements of these two different forecasting methods, the Absolute Error (MAE), Mean Square Error (MSE), and Root Mean Square Error (RMSE) forecasting error performance measures used in the study are briefly summarized.

3.2.1 ARIMA Models for Classical Time Series Model

The autoregressive integrated moving average (ARIMA) model, which is frequently used in classical linear time series models consists of two components: autoregressive AR(p) and moving average MA(q). The AR(p) component models the relationship between a series' lag length and its past values, and the MA(q) component models the relationship between a series' lag length and the error squares. The model could be written as follows (Brooks 2019: 351–352):

$$\phi(L)y_t = \mu + \theta(L)u_t \tag{1}$$

where,

$$\phi(L) = 1 - \phi_1 L - \phi_2 L^2 - \cdots - \phi_p L^p \tag{2}$$

$$\phi(L) = 1 + \theta_1 L + \theta_2 L^2 + \cdots + \theta_q L^q \tag{3}$$

or

$$y_t = \mu + \phi_1 y_{t-1} + \phi_2 y_{t-2} + \cdots + \phi_p y_{t-p} + \theta_1 u_{t-1} + \theta_2 u_{t-2} + \cdots + \theta_q u_{t-q} + u_t \tag{4}$$

with

$$E(u_t) = 0;\ E\left(u_t^2\right) = \sigma^2;\ E(u_t u_s) = 0,\ t \neq s \tag{5}$$

Equation 5, assumes that the mean of the error term is randomly distributed around zero, its variance does not change over time, and there is no serial correlation between the error terms in different periods.

In the selection of ARIMA models, the Box-Jenkins approach is the most widely used method. The two criteria suggested by Box and Jenkins (1976) for model selection are that the model has the least parameters and that the translatability feature is realized (Enders 2014: 78). The Box and Jenkins approach consists of three stages: definition, estimation, and diagnostic (Brooks 2019; Box 2013). In the first step, this determines the lag orders for the AR(p) and MA(q) required to capture the dynamic properties of the data. When determining the lag orders, the coefficients of the ARIMA model must be statistically significant. In the second step, the parameters of the ARIMA (p, q) model defined in step 1 are estimated by an estimation method known as ordinary least squares (OLS) or maximum likelihood (ML). The third step involves model checking, that is, determining whether the model specified and estimated is adequate. The residuals of the estimated ARIMA (p, q) model are checked for serial correlation and heteroscedasticity using the LM and White tests, respectively. If these problems are detected, it indicates that the initially specified model is inadequate to capture the characteristics of the series.

3.2.2 Artificial Neural Networks (ANN) Model for Machine Learning Method

The human brain is an extremely complex and multi-layered system that processes information through the interaction between neurons, which are nerve cells (Tozzi 2019). Despite today's scientific developments, the working mechanism of the human brain has not been fully resolved by scientists. The process of learning information in the human brain is shaped not only by neurons but also by the continuous and multifaceted interactions between these nerve cells (Pessoa 2014). The human brain was taken as a source of inspiration when developing artificial neural network models. The learning processes and data processing of artificial neural networks with the machine learning method have become the basis of this method. Artificial neural network research can be examined under three main headings: (Cheng and Titterington 1994; Allende et al. 2002; Gevrey et al. 2003): The first field includes experimental studies based on the science of biology, physiology, and molecular biology. In this field, attempts are made to understand the basic functioning of artificial neural networks and the interaction processes between artificial nerve cells. The second field includes engineering applications inspired by the human brain structure calculation style. In this area, information is distributed as an analog pattern between the signal neurons and enables the learning process by highlighting their computational capabilities. The third area examines the mathematical basis of neurocalculus, which reveals the basic principles of neuronal cells that have learning capabilities. The field of statistics is particularly closely related to the field of application of artificial neural networks and has led to the development of many application areas such as image recognition and time series analysis. These three areas reveal the multifaceted nature of artificial neural networks and the interaction between different disciplines (Prieto et al. 2016).

The ANN model, a machine learning method, is an alternative prediction method that can explain multivariate and nonlinear relationships. The accurate and successful predictions of the ANN model are based on statistical methods that form the basis of the design of neurocomputers that simulate the behavior of artificial neurons. ANN models provide a theoretical framework for neurocomputers. Therefore, statistical learning is important in improving data analysis and prediction success (Almeida

2002). In addition, these models are effective in addressing a wide range of challenges, including data processing, classification, and nonlinear time series prediction (Baxt 1990; Pirovano and Heringa 2010). ARIMA models are frequently used by econometricians to forecast financial time series. However, the presence of non-linear and chaotic structures in financial time series has led to the increased use of ANN models, one of the machine learning methods, as an alternative to ARIMA models in recent years (Pang et al. 2020).

An artificial neuron consists of five parts: inputs, weights, summation function-activation function, and outputs. Inputs are the data coming to neurons. The data coming from these inputs can process information. Weights are transmitted to the core by multiplying the information coming to the artificial neuron with the weights of the connections they come from before reaching the core via the inputs. The summation function is a function that calculates the net input of that cell by adding the inputs coming to an artificial neuron multiplied by the weights. The activation function is a function that takes the weighted sum of all inputs in the previous layer and then produces an output value and passes it to the next layer. Outputs The value coming out of the activation function is the output value of the cell. These five parts together direct the success of ANN models in data processing, modeling, and prediction tasks (Basheer and Hajmeer 2000).

In this study, a feedforward supervised learning algorithm with two hidden layers was used for estimating the ANN model. The number of hidden layers in the ANN model is determined according to the characteristics of the dataset under study. Nonlinear relationships observed in the structure of the data in the ANN model may affect the learning capacity of this model (Goh 1995). ANN models, also known as Black-Box models, one of the machine learning methods, do not have an assumption that requires prior knowledge about the structural and functional properties of the system under study (Guidotti et al. 2018). Instead, these models learn the relationships between inputs and outputs autonomously by analyzing past data values. Input and output layers are called 'hidden layers' because they contain units that process information but cannot be directly observed. The number of input units usually depends on the characteristics of the data and the requirements of the application at hand (Stathakis 2009). Due to these features, ANN models are preferred in analyses due to their success in revealing complex and non-linear relationships (Ahmad et al. 2017; Pashaei and Pashaei 2021).

In the literature, various artificial neural network strategies have been developed; these strategies include feedforward networks, Hopfield networks, self-organizing maps, and radial basis networks (Barreto and Araujo 2004). Feedforward neural networks stand out as the most common and general type of neural network due to their simple structure and wide usage area; these networks have been effectively used in many problems where nonlinear relationships need to be learned. This study is based on the feedforward neural network structure in the modeling process and the backpropagation algorithm, a popular learning algorithm to optimize the network.

3.2.3 Comparison of Forecasting Performance of Models

This study uses three different criteria, namely Mean Absolute Error (MAE), Mean Square Error (MSE), and Root Mean Square Error (RMSE) to compare the performance efficiency of the ARIMA and the ANN models in forecasting the USD/TRY exchange rate. These performance measures are used to compare alternative forecasting models. Among the alternative models, the model with the smallest performance measures is chosen as the most successful model for prediction. Forecasting performance measures are used to determine how successful the model is in predicting or how small the prediction errors are. The forecasting performance measure errors used in this study are explained as follows (Brooks 2019):

- Mean Absolute Error (MAE): MAE is the average absolute difference between the observed and predicted values. It shows that the model with the smallest MAE value among the alternative models is more successful in predictions. The MAE formula is:

$$MAE = \frac{1}{n}\sum_{i=1}^{n}|y_i - \hat{y}_i| \tag{6}$$

- Mean Square Error (MSE): MSE calculates the difference between predicted and true values. It is a more sensitive measure because it allows for greater penalties for large errors. The formula for MSE is:

$$MSE = \frac{1}{n}\sum_{i=1}^{n}(y_i - \hat{y}_i)^2 \tag{7}$$

- Root Mean Square Error (RMSE): RMSE is calculated by taking the square root of MSE. RMSE is used to understand how much the estimates deviate and provides errors at the same scale as the estimated unit. Smaller RMSE means better performance.

$$RMSE = \sqrt{MSE} \tag{8}$$

4 Empirical Findings

In this study, the natural logarithm of the ER variable is used for analysis. To make reliable predictions in the ARIMA modeling approach, it is a basic requirement that the analyzed time series be stationary. Therefore, the Augmented Dickey-Fuller (ADF 1979) and Phillips-Perron (PP 1988) unit root tests are applied to evaluate the stationarity of the LER variable. Table 1 presents the results of the unit root tests for the LER variable. The ADF and PP unit root tests test the null hypothesis of the existence of a unit root against the alternative hypothesis that the series is stationary, respectively. According to the results presented in Table 1, it is concluded that the null hypothesis is rejected according to the 1% probability value of the first difference of the LER variable. In other words, the LER variable is determined to be a first difference stationary series.

To convert a non-stationary LER variable, exchange rate returns (DLER) are calculated as follows:

$$DLER = \log(ER_t) - \log(ER_{t-1}) = \log\left(\frac{ER_t}{ER_{t-1}}\right) \tag{9}$$

The most appropriate ARMA (p,q) model for the DLER is estimated and the results are shown in Table 2.

According to the estimation results, the most appropriate model selected by the Akaike information criterion (AIC) is the ARMA (1,2) model. There is no serial correlation problem in the residuals of the ARMA (1,2) model, but the ARCH (1) effect is detected. Since the volatility of the DLER variable will not be modeled in the study and only intra-period forecast values will be obtained, the ARCH (1) effect is not taken into account in the analyses.

Figure 1 shows a graph showing the intra-period forecast values and actual values obtained from the ARIMA model. The closer the observed values and the estimated

Table 1 Unit root tests for the LER variable

Series	ADF test		PP test	
	Level	First difference	Level	First difference
LER	− 1.2290 (2)	− 9.2891 (1)***	− 1.0994 (7)	− 8.4046 (10)***

Note The values in parentheses represent the optimum lag length by the SIC; ***$p < 0.01$, **$p < 0.05$, **$p < 0.10$

Table 2 ARMA (1,2) model estimation results

Dependent variable: DLER	Coefficient	t-statistics/Prob.
Intercept	0.0234	5.3549 (0.0000)***
AR(1)	0.9713	59.7894 (0.0000)***
MA(1)	− 0.4789	− 7.2819 (0.0000)***
MA(2)	− 0.5108	− 7.7616 (0.0000)***
Diagnostic Tests	Test statistics	Prob.
Serial Correlation LM (12) test	8.9976	(0.7031)
ARCH (1)	10.2290	(0.0014)***
$R^2 = 0.2287$		

Note *** $p \leq 0.01$

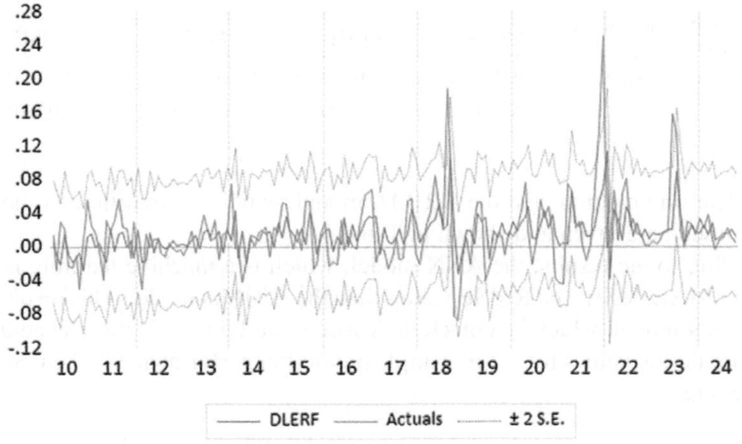

Fig. 1 The DLER forecast (DLERF) and actuals values (DLER)

values are to each other, the model is considered successful. This result indicates that the in-sample prediction values of the ARMA (1,2) model were made successfully and accurately. However, to evaluate whether the in-sample prediction values of the estimated ARMA (1,2) model are successful, it is necessary to compare the prediction error performance measures of the ANN model estimated with machine learning, which is an alternative prediction method.

In this study, the ANN model, being one of the machine learning methods representing an alternative method to linear time series analysis, was estimated using R software. First of all, from the data set of the USD/TRY exchange rate, the minimum and maximum values were obtained. Then, to increase the prediction success and accuracy of the ANN model, all the input and output parameters were normalized and scaled to take values between 0 and 1. These scaled values (SV) are obtained as follows:

$$SV = \frac{DLER_t - DLER_{\min}}{DLER_{\max} - DLER_{\min}} \tag{10}$$

Scaling the input and output parameters of the ANN model allows for standardization and more effective and precise predictions. In the third stage, the observed values were divided into two parts: the training and test data. The first 160 observations in the dataset were assigned to the training data set, while the last 17 were allotted to the test data set. Fourthly, the ANN model was estimated by using the values in the training data set, and intra-term prediction values were obtained. This procedure enables the ANN model to learn from the training data and subsequently make predictions. Then, the performances of the ANN model predictions obtained from the test data and training data were compared with the prediction error criteria. Finally, the prediction error performance criteria of the ANN model, which is a

Table 3 Comparision of the prediction performances of ARIMA and ANN models

Model	MAE	MSE	RMSE
ARIMA	0.027375	0.001624	0.040311
ANN	0.017943	0.000628	0.025069

machine learning method, and the ARIMA model, which is a traditional time series analysis model, are given and shown in Table 3.

According to the results, the ANN model, which is a machine learning method, gave lower MAE, MSE, and RMSE values than the ARIMA model. This result shows that the ANN model, which is a machine learning method, gives more accurate and reliable prediction values for the in-sample prediction performance of the USD/TRY exchange rate.

5 Conclusion

This study aims to obtain the intra-period forecast values of the USD/TRY exchange rate using monthly data for Turkey from January 2010 to October 2024 using two alternative models. For this purpose, the ARIMA model, one of the linear time series analysis methods for the USD/TRY exchange rate, and the ANN model, one of the machine learning methods, were preferred. The performance of these two alternative models was compared with the Mean Absolute Error (MAE), Mean Square Error (MSE), and Root Mean Square Error (RMSE) forecast error measurement values and the model that gives the smallest forecast error measurement value was preferred.

The findings show that the ANN model, one of the machine learning methods, gives a lower forecast error value than the ARIMA model, one of the linear time series analysis methods. This result shows that the ANN model gives more accurate intra-period forecast values of the USD/TRY exchange rate. In addition, it is concluded that the ANN model makes more successful forecasts when exchange rate movements are volatile. In addition, it is concluded that the ANN model makes more successful forecasts when exchange rate movements are volatile. ANN models are successful in capturing nonlinear and chaotic structures in financial time series such as exchange rates. The findings in this study are supported by previous studies such as Panda and Narasimhan (2007), Kadılar et al. (2009), Pradhan and Kumar (2010), Aydın and Cavdar (2015), and Sun et al. (2020). Studies have shown that ANN models provide more reliable results than traditional linear time series models in periods of financial crisis and volatility of exchange rates.

Sudden fluctuations in exchange rates have an immediate impact on financial markets, whereas their macroeconomic effects are often observed over a longer term. Increased volatility in the USD/TRY exchange rate can adversely affect both companies engaged in international trade and domestic investors. For example, an increase in the USD/TRY exchange rate increases costs for companies that depend on raw materials and intermediate goods. Therefore, it is important for companies that do business with foreign currencies to accurately predict exchange rate price movements. If companies can accurately predict current and future input costs, they can also manage their risks. In addition, exchange rate forecasts play an important role in determining monetary and fiscal policies by central banks and policymakers. Successful exchange rate forecasts contribute to the development of policies that can have an impact on inflation rates and economic growth.

Finally, for new studies on economic time series such as exchange rates, in-sample or out-of-sample predictions can be made with different frequencies such as real-time, daily, weekly, or monthly using machine learning methods. Hybrid ARCH family models can be used as alternative prediction methods to predict volatility movements of financial time series. In addition, alternative prediction models are estimated using various machine learning algorithm methods and prediction error measures can be used to compare the performances of these models. As such, with the use of other methods such as support vector machines, decision trees, random forests, and deep learning models, it is possible not only to forecast different frequency financial time series like daily and weekly but also to measure and compare the effectiveness of alternative methods.

References

Ahmad MW, Mourshed M, Rezgui Y (2017) Trees versus neurons: comparison between random forest and ANN for high-resolution prediction of building energy consumption. Energy Build 147:77–89

Allende H, Moraga C, Salas R (2002) Artificial neural networks in time series forecasting: a comparative analysis. Kybernetika 38:685–707

Almeida JS (2002) Predictive non-linear modelling of complex data by artificial neural networks. Curr Opin Biotechnol 13(1):72–76

Aydın AD, Cavdar SC (2015) Comparison of prediction performances of artificial neural network (ANN) and vector autoregressive (VAR) models by using the macroeconomic variables of gold prices, borsa Istanbul (BIST) 100 index and US dollar-Turkish Lira (USD/TRY) exchange rates. Procedia Econ Finance 30:3–14

Barreto GA, Araujo AFR (2004) Identification and control of dynamical systems using the self-organizing map. IEEE Trans Neural Networks 15(5):1244–1259

Basheer IA, Hajmeer M (2000) Artificial neural networks: fundamentals, computing, design, and application. J Microbiol Methods 43(1):3–31

Baxt WG (1990) Use of an artificial neural network for data analysis in clinical decision-making: the diagnosis of acute coronary occlusion. Neural Comput 2(4):480–489

Box GEP, Jenkins GM (1976) Time series analysis: forecasting and control, Revised. Holden-Day, CA

Box G (2013) Box and Jenkins: time series analysis, forecasting and control. In: A Very British Affair. Palgrave Advanced Texts in Econometrics. Palgrave Macmillan, London

Brooks C (1998) Linear and non-linear (non-) forecastability of high-frequency exchange rates. J Forecast 16:125–145

Brooks C (2019) Introductory econometrics for finance. Cambridge University Press, United Kingdom

Carol A (2009) Market risk analysis, value at risk models. Wiley

Charef F (2024) Exchange rate forecasting: nonlinear GARCH-NN modeling approach. Ann Data Sci 11(3):947–957

Chen W, Xu H, Jia L, Gao Y (2021) Machine learning model for Bitcoin exchange rate prediction using economic and technology determinants. Int J Forecast 37(1):28–43

Cheng B, Titterington DM (1994) Neural networks: a review from a statistical perspective. Stat Sci 9(1):2–30

Clements MP, Hendry DF (2000) Forecasting economic tim series. Cambridge University Press, United Kingdom

Dickey DA, Fuller WA (1979) Distribution of the estimators for autoregressive time series with a unit root. J Am Stat Assoc 74(366a):427–431

Enders W (2014) Applied econometric time series. Wiley, New York

Faini R, Melo J (1990) Adjustment, investment and the real exchange rate in developing countries. Economic Policy 5(11):491–519

Franses PH, Homelen PV (1998) On forecasting exchange rates using neural networks. Appl Financ Econ 8(6):589–596

Gevrey M, Dimopoulos I, Lek S (2003) Review and comparison of methods to study the contribution of variables in artificial neural network models. Ecol Model 160(3):249–264

Goh ATC (1995) Back-propagation neural networks for modeling complex systems. Artif Intell Eng 9(3):143–151

Guidotti R, Monreale A, Ruggieri S, Turini F, Giannotti F, Pedreschi D (2018) A survey of methods for explaining black box models. ACM Comput Surv 51(5):1–42

Hau H, Rey H (2005) Exchange rates, equity prices, and capital flows. Rev Financ Stud 19(1):273–317

Kadılar C, Şimşek M, Aladağ ÇH (2009) Forecasting the exchange rate series with ANN: the case of Turkey. Ekonometri Ve İstatistik 9:17–29

Kamruzzaman J, Sarker RA (2004) ANN-Based forecasting of foreign currency exchange rates. Neural Inform Process-Lett Rev 3(2):49–58

Kristjanpoller W, Hernandez E (2017) Volatility of main metals forecasted by a hybrid ANN-GARCH model with regressors. Expert Syst Appl 84:290–300

Lane PR, Ferretti GMM (2002) External wealth, the trade balance, and the real exchange rate. Eur Econ Rev 46(6):1049–1071

Levich RM (1985) Empirical studies of exchange rates: price behavior, rate determination and market efficiency. Handb Int Econ 2:979–1040

McKenzie MD (2002) The impact of exchange rate volatility on international trade flows. J Econ Surv 13(1):71–106

Moosa IA (2000) Exchange rate forecasting: techniques and applications. Palgrave Macmillan

Panda C, Narasimhan V (2007) Forecasting exchange rate better with artificial neural network. J Policy Model 29(2):227–236

Pang X, Zhou Y, Wang P, Lin W, Chang V (2020) An innovative neural network approach for stock market prediction. J Supercomput 76:2098–2118

Papaioannou MG (2006) Exchange rate risk measurement and management: Issues and Approaches for firms. IMF Working Paper No. 06/255, Available at SSRN: https://papers.ssrn.com/sol3/papers.cfm?abstract_id=947372

Pashaei E, Pashaei E (2021) Training feedforward neural network using enhanced black hole algorithm: a case study on COVID-10 related ACE2 gene expression classification. Arab J Sci Eng 46:3807–3828

Pessoa L (2014) Understanding brain networks and brain organization. Phys Life Rev 11(3):400–435

Phillips PCB, Perron P (1988) Testing for a unit root in time series regression. Biometrika 75(2):335–346

Pirovano W, Heringa J (2010) Data mining techniques for the life sciences. Humana Press, United States

Pradhan RP, Kumar R (2010) Forecasting Exchange rate in India: an application of artificial neural network model. J Math Res 2(4):111–117

Prieto A, Prieto B, Ortigosa EM, Ros E, Pelayo F, Ortega J, Rojas I (2016) Neural networks: an overview of early research, current frameworks and new challenges. Neurocomputing 214:242–268

Spyros M, Nikolas B (2016) Forecasting and uncertainty: a survey. Risk Decis Anal 6(1):37–64

Stathakis D (2009) How many hidden layers and nodes? Int J Remote Sens 30(8):2133–2147

Sun S, Wang S, Wei Y (2020) A new ensemble deep learning approach for exchange rates forecasting and trading. Adv Eng Inform 46:1–10

Tozzi A (2019) The multidimensional brain. Phys Life Rev 31:86–103

Yavuz NÇ (2014) Finansal Ekonometri. Der Yayınları, İstanbul

Zhang GP (2003) Time series forecasting using a hybrid ARIMA and neural network model. Neurocomputing 50:159–175

The Imbalanced Data Problem: Investigating Factors Affecting Financial Freedom Using Data Mining Techniques with SMOTE Method

Abdurrahman Coşkuner and Ömer Faruk Rençber

Abstract Class distribution disparities in datasets often result in imbalanced data issues, which can significantly impact model performance. This study investigates the effects of such imbalances on the performance of XGBoost and Support Vector Machines (SVM), specifically in the context of a five-class classification problem using the financial freedom index as the target variable. Initially, both models were applied to the imbalanced dataset, highlighting the performance degradation caused by the data imbalance. To mitigate this issue, the Synthetic Minority Oversampling Technique (SMOTE) was employed to generate a balanced dataset, after which the models were re-evaluated. Comparative analysis revealed that the XGBoost algorithm demonstrated superior performance relative to the SVM method once the data imbalance was addressed. Moreover, the improvement in classification accuracy for XGBoost was notably higher compared to SVM following the application of the SMOTE technique, underscoring the robustness of XGBoost in handling imbalanced data.

Keywords Financial freedom · XGBOOST · Support vector · SMOTE

1 Introduction

The escalating complexity and sheer volume of data in recent years have amplified the significance of advanced machine-learning techniques across a wide range of disciplines. One of the critical obstacles in classification tasks arises from imbalanced datasets, where underrepresentation of certain classes can severely degrade model

A. Coşkuner (✉)
Faculty of Economics and Administrative Sciences, Gaziantep University, Gaziantep, Türkiye
e-mail: acoskuner02@gmail.com

Ö. F. Rençber
Faculty of Economics and Administrative Sciences, Gaziantep University, Gaziantep, Türkiye
e-mail: dr.ofrencber@gaziantep.edu.tr

M. Gün and B. Kartal (eds.), *Machine Learning in Finance*, Contributions to Finance and Accounting, https://doi.org/10.1007/978-3-031-83266-6_6

performance. Addressing these challenges is fundamental for enhancing the accuracy and dependability of predictive models. Consequently, gaining a comprehensive understanding of various learning techniques and their methodological approaches is essential for researchers and practitioners aiming to optimize model performance and effectively manage the complexities presented by real-world data scenarios.

Machine learning techniques are generally categorized into supervised, unsupervised, and reinforcement learning methods. In data mining applications involving supervised and reinforcement learning, datasets are typically divided into three subsets: training, testing, and validation sets. The training set generally comprises 70–80% of the dataset, while the test set accounts for 20–30% of the remaining data (Yoo et al. 2012). The training set is used to train the algorithm, and the performance of the trained model is subsequently assessed using the test set (Neelamegam and Ramaraj 2013). However, when dealing with datasets containing multiple classes, irregularities can arise, leading to a reduced probability of adequately representing minority classes in the training set. This imbalance may result in the algorithm underlearning, thereby decreasing its overall classification accuracy (Agrawal et al. 2015). The issue of class imbalance has recently been acknowledged as a significant challenge in data mining research (Gu et al. 2022). To address this challenge, various strategies have been developed, including Data-Level Approaches, Algorithm-Level Approaches, Feature-Based Approaches, and Hybrid Approaches (Yadav and Bhole 2020). Data-level approaches specifically focus on mitigating the class imbalance by modifying the dataset, often by balancing the distribution of instances among classes (Fernández et al. 2018). One of the most widely used data-level techniques is the Synthetic Minority Oversampling Technique (SMOTE), introduced by Chawla et al. (2002). The SMOTE method combats data imbalance by generating synthetic samples to over-sample the minority class, thereby enhancing the representation of minority instances and improving model performance (Czarnowski 2022).

The present study aims to illustrate the impact of imbalanced data on data mining methods by applying XGBoost and Support Vector Machines (SVM) to the Financial Freedom Index 2022 dataset. The Financial Freedom Index evaluates countries' economic freedom based on four main pillars: rule of law, government size, regulatory efficiency, and open markets (Heritage 2023). XGBoost, known for its parallel tree-boosting approach, is celebrated for its speed and accuracy, making it a powerful solution for a variety of data science challenges (Zhang and Zhan 2017). In contrast, SVM is designed to optimize the objective function globally during training, which reduces the risk of overfitting. This capability allows SVM to effectively handle large feature spaces and manage high-dimensional data (Shen et al. 2007). Both methods have unique strengths attributed to their distinct advantages. Understanding how these popular and effective algorithms perform in the presence of imbalanced data distributions is of significant interest and adds valuable insights into their robustness and applicability in real-world scenarios.

In the application phase, five class classifications were initially performed using both XGBoost and Support Vector Machines (SVM) on the Financial Freedom Index dataset. It was observed that the classifications generated by both methods revealed an imbalanced data distribution. The accuracy rate for XGBoost was recorded at

53%, while SVM achieved an accuracy rate of 59%. To address the imbalance, the Synthetic Minority Oversampling Technique (SMOTE) was applied to create a balanced dataset. Subsequent application of both methods to this balanced dataset demonstrated an improvement in classification accuracy, with XGBoost achieving a 79% accuracy rate and SVM reaching 69%. The results indicate a substantial improvement for both methods, with XGBoost's accuracy increasing by 26% and SVM's by 10% following data balancing. These findings highlight the superior performance of XGBoost over SVM in handling imbalanced data, as well as the significant impact of using techniques like SMOTE to enhance model accuracy. The study underscores the critical role of data-balancing strategies in improving the performance of machine learning models, with XGBoost benefiting the most from such adjustments.

The structure of the study is as follows: the first section provides an introduction, followed by a literature review in the second section. The third section describes the XGBoost and SVM methods, while the fourth section explains the SMOTE technique. The fifth section offers insights into the Economic Freedom Index, which serves as the data source for the study. The sixth section details the application process, and the final section presents the results and recommendations.

2 Literature Review

Numerous studies have demonstrated the superior performance of XGBoost compared to other popular methods, such as Random Forest and Support Vector Machines (SVM). For instance, Gao et al. (2020) highlighted that XGBoost consistently outperformed other ensemble learning models, particularly in classification tasks involving imbalanced datasets. This advantage makes XGBoost a highly suitable choice for domains where data imbalance is prevalent. Furthermore, Atsauri et al. (2023) reported that optimizing XGBoost specifically for imbalanced datasets led to a marked improvement in performance, emphasizing the algorithm's adaptability and effectiveness.

Support Vector Machines (SVM) are also highly regarded in machine learning for their accuracy and versatility in classification tasks, and they are widely applied across various fields. The algorithm's effectiveness has been well-documented in disciplines such as medicine, bioinformatics, and finance. For example, Lin et al. (2022) demonstrated the potential of SVMs in healthcare by predicting hospital length of stay based on electronic medical records (EMR) at the time of admission. This study underscored SVM's capability to handle complex medical data and provide valuable insights to healthcare providers. In a related study, Njoroge et al. (2006) found that SVMs yielded higher accuracy in classifying cervical cancer cells using Fourier Transform Infrared (FTIR) spectroscopy data compared to the traditional Pap test. Specifically, the SVM model achieved a classification accuracy of 72%, significantly outperforming the Pap test's 43% accuracy rate.

In the field of bioinformatics, Support Vector Machines (SVMs) have emerged as a prominent classification tool, particularly in subfields such as genomics and proteomics. Zhao et al. (2018) underscored the importance of SVMs in identifying molecular signatures for tumor classification, illustrating their effectiveness in detecting cancer biomarkers. Additionally, Cui et al. (2014) leveraged SVMs for predicting transcription factor target genes, demonstrating the algorithm's utility in exploring gene regulation mechanisms. Similarly, Pertea et al. (2007) showcased the superior performance of SVMs over other classification methods when analyzing biological sequence data, highlighting their robustness in bioinformatics applications.

SVMs have also shown considerable success in protein classification and structure analysis within computational biology. Akutsu et al. (2012) emphasized the significance of SVMs in understanding protein function and evolution, using the algorithm to identify conserved regions within protein structures. Ben-Hur et al. (2008) provided a comprehensive review of various SVM applications in computational biology, noting the algorithm's high accuracy and its ability to handle complex datasets effectively. Moreover, Xia et al. (2008) demonstrated the success of SVMs in differentiating structural exons from alternatively spliced exons, employing the algorithm to predict features associated with alternative splicing.

The application of Support Vector Machines (SVMs) extends beyond the biological and medical fields, demonstrating their versatility in financial analysis as well. Jiang et al. (2009) developed an SVM-based classification model to assess the financial performance of companies, highlighting the algorithm's adaptability to financial data analysis. This underscores the broader applicability of SVMs, illustrating that they are not confined to disciplines like bioinformatics and healthcare but can also serve as a robust classification tool in areas such as finance.

In the health domain, XGBoost has also proven to be highly effective. Abdeltawab et al. (2022) demonstrated the algorithm's superior performance compared to other classifiers by employing XGBoost to predict respiratory support requirements for COVID-19 patients, emphasizing its robustness and predictive accuracy in medical contexts. Furthermore, Gu et al. (2021) reported that XGBoost consistently delivers high accuracy across various datasets, particularly when combined with techniques designed to prevent overfitting. This makes XGBoost a reliable and effective model for diverse data-driven applications.

The problem of imbalanced data often poses a significant challenge to the success of classification algorithms. Literature highlights the effectiveness of the Synthetic Minority Oversampling Technique (SMOTE) in improving algorithm performance in such scenarios. He et al. (2021) applied SMOTE in combination with XGBoost, Random Forest, Support Vector Machine (SVM), and logistic regression to develop an interpretable and practical model for ship arrest decisions, with SMOTE-XGBoost achieving the best results. Similarly, Ahsan et al. (2018) explored the impact of SMOTE on the performance of XGBoost, Random Forest, and SVM for cybersecurity purposes using phishing data. Their findings revealed that XGBoost when combined with SMOTE, had a higher success rate compared to other methods.

Wang et al. (2020) employed the XGBoost algorithm to classify diesel fuel brands, addressing data imbalance by applying SMOTE to balance class distributions. This application led to a 19.33% increase in the model's accuracy. Hasib et al. (2022) developed a predictive model for student success in secondary education using various classification techniques, including logistic regression, k-nearest neighbors, SVM, XGBoost, and Naive Bayes. They also used SMOTE to balance the data and found that the SVM method yielded the highest accuracy.

Comparative studies between XGBoost and Support Vector Machine (SVM) methods have generally shown the superior performance of XGBoost. Mateo et al. (2021) applied various machine learning algorithms to classify flower honey, finding that XGBoost outperformed other methods, whereas SVM demonstrated relatively lower performance. Similarly, Yusri et al. (2022) evaluated XGBoost and SVM for classifying water quality using a four-class dataset. Their results indicated that XGBoost achieved higher classification accuracy compared to SVM. In another study aimed at developing an electronic nose system, Binson et al. (2021) compared the two methods for classifying exhalation samples from healthy individuals and patients with lung cancer, COPD, and asthma. The findings showed that XGBoost provided a more accurate classification rate than SVM.

Studies in the literature highlight the prominence of XGBoost and Support Vector Machines (SVM) as leading algorithms for classification tasks. XGBoost is particularly noted for its exceptional performance, scalability, and algorithmic optimization, making it well-suited for high-dimensional and complex datasets. Its superior efficiency has been demonstrated across various data science and medical applications. On the other hand, SVM is renowned for its high accuracy and versatility, successfully handling diverse data types in fields such as healthcare, bioinformatics, and finance.

Moreover, XGBoost's performance on imbalanced datasets is significantly improved with data-balancing techniques like SMOTE. Comparative analyses across different domains often show that XGBoost achieves higher success rates. Nonetheless, SVM remains a robust alternative, offering flexibility and accuracy in numerous applications. Thus, the choice between these algorithms largely depends on the data structure, application area, and desired performance outcomes. The subsequent sections of this study provide detailed information about these two methods.

3 XGBoost and Support Vector Machines

Decision trees are structured on a hierarchical, tree-like framework, enabling predictions on an output variable based on a series of defined rules (Torlay et al. 2017). eXtreme Gradient Boosting (XGBoost) is an ensemble learning method that combines multiple decision trees with initially low classification performance, iteratively improving to construct highly efficient models. XGBoost outperforms traditional Gradient Boosting Decision Tree (GBDT) methods in terms of speed, scalability, and generalization (Chen and Guestrin 2016). Additionally, it effectively addresses overfitting and bias issues through advanced hyperparameter tuning (Morde 2019). These features contribute to XGBoost's widespread adoption and preference in various applications.

Support Vector Machines (SVM) are another essential machine learning approach, applicable to both classification and regression tasks (Cervantes et al. 2020). SVM operates by constructing a decision function in the form of a hyperplane that optimally separates observations from different classes based on the information in the dataset. This method is formulated as an optimization problem that seeks to maximize the margin of the hyperplane, thereby enhancing classification accuracy (Pisner and Schnyer 2020).

Support Vector Machines (SVM) and other classification algorithms have two primary objectives. The first is to maximize the accuracy of the classifier, which means optimizing the model to correctly label new instances it encounters, thereby improving its overall performance. The second objective is to ensure the generalizability of the classifier. This involves making sure the model performs consistently well on new, unseen data, and optimizing its ability to repeat its performance under different conditions. This section has provided an overview of the two key methods used in the study: XGBoost and Support Vector Machines. The following section will discuss the SMOTE method, which was employed to address the issue of imbalanced data.

4 Synthetic Minority Over-Sampling Technique (SMOTE)

The uneven distribution of classes within a dataset often results in imbalanced data, posing a challenge for algorithms to accurately classify minority class instances. This imbalance occurs when classes with fewer instances represent only a small proportion compared to the majority class, reducing the algorithm's ability to learn effectively. To address this, Chawla et al. (2002) introduced the Synthetic Minority Over-Sampling Technique (SMOTE), which oversamples the minority class by generating synthetic examples. SMOTE works by selecting an instance from the minority class and finding its nearest neighbors to identify similar examples. Synthetic samples are then created by randomly choosing from these similar instances and generating new data points.

The method operates through a series of steps (Fernández et al. 2018). First, the total number of samples, N (an integer), is determined to achieve an approximate 1:1 class distribution. An iterative process is then carried out: a sample from the minority class is randomly selected from the training set, and its K nearest neighbors (typically $K = 5$) are identified. N synthetic samples are generated by randomly selecting instances from these K neighbors and adding them to the dataset through augmentation. SMOTE uses examples (denoted as x and xR) selected from the minority class based on a similarity measure. The synthetic examples are produced through a linear combination of these original instances, expressed in a mathematical formulation as described by Blagus and Lusa (2013).

The formula for generating synthetic samples using SMOTE is given by:

$$s = x + u.(x^R - x) \quad (1)$$

where u is a randomly chosen number between 0 and 1, and x^R is randomly selected from the nearest neighbors of x within the minority class.

The Synthetic Minority Over-Sampling Technique (SMOTE) is a highly effective method for addressing class imbalance issues. By generating synthetic examples of the minority class, SMOTE enhances the representational power of underrepresented data, significantly improving the performance of classification algorithms on imbalanced datasets. Developed by Chawla et al. (2002), SMOTE creates synthetic data points based on the similarity of selected examples and their neighbors in the minority class. This approach ensures a more balanced class distribution, enabling classification models to better recognize and classify minority instances, thus enhancing generalization performance.

This section has outlined the SMOTE method employed in this study to address class imbalance. The next section will provide details about the Economic Freedom Index, which serves as the dataset for this research.

5 Economic Freedom Index

The Economic Freedom Index (EFI), developed by the Heritage Foundation in collaboration with the Wall Street Journal, is a crucial instrument for global economic freedom assessment. Since its inception in 1995, the index has offered a comprehensive evaluation of economic freedom, encompassing components such as property rights, government integrity, and business freedom. This multidimensional approach provides a detailed analysis of how economic policies and institutional frameworks impact overall economic performance and social welfare (Erilli 2018; Evans and Naurodski 2019). Studies have consistently demonstrated a strong correlation between economic freedom and favorable economic outcomes. Specifically, higher levels of economic freedom are linked to increased prosperity, improved social conditions, and enhanced human development (Gwartney et al. 2022; Naanwaab 2013; Vitenu-Sackey 2022).

The index's twelve components offer a robust framework for analyzing the influence of economic freedom on various macroeconomic indicators, including foreign direct investment (FDI) and productivity (Naanwaab 2013). Nevertheless, the EFI's methodology has sparked academic debates. Some scholars argue that the overall index may not accurately capture the economic realities of certain countries, while others emphasize the need to analyze individual components to gain a more nuanced understanding of economic dynamics (Belanová et al. 2023; Brkić et al. 2020). Notably, sub-indices such as trade freedom and investment freedom have been identified as significant determinants of economic outcomes (Li 2015). This underscores the necessity for further empirical research to explore the distinct effects of these components on economic growth and development, offering valuable insights for policymakers seeking to enhance economic performance (Brkić et al. 2020; Yevdokimov et al. 2018).

Beyond economic impacts, EFI has been associated with broader social outcomes, including democracy and poverty alleviation. High levels of economic freedom are often correlated with greater political stability and improved governance, creating a conducive environment for sustainable economic growth (Vitenu-Sackey 2022; Yevdokimov et al. 2018). Consequently, the EFI serves as a critical indicator of economic conditions institutional quality, and governance structures across nations (Ott 2018).

The EFI evaluates 177 countries based on their economic freedoms across 12 categories, ranking them according to their respective scores. This index serves as a valuable resource for both domestic and foreign investors, providing essential information to guide investment decisions. It also offers insights for policymakers, aiding in the formulation of policies and reforms aimed at enhancing economic performance. The 12 categories are grouped under four main headings, each scored on a scale of 0–100. The overall score for a country is calculated by averaging the scores across these categories, and countries are subsequently ranked based on these averages (HERITAGE 2023).

The four main headings and their components are as follows:

- Rule of Law: This includes Property Rights, Judicial Effectiveness, and Government Integrity.
- Government Size: This category measures Government Expenditure, Tax Burden, and Fiscal Health.
- Regulatory Efficiency: It encompasses Business Freedom, Monetary Freedom, and Labor Freedom.
- Open Markets: This consists of Financial Freedom, Trade Freedom, and Investment Freedom.

The rule of law component assesses the extent to which laws are effectively applied, the protection of property rights, and the treatment of individuals within the justice system. The government dimension evaluates the level of government intervention in the economy, including public spending and tax policies. Regulatory efficiency measures how easily businesses can be established and operated, as well as the complexity of regulatory processes. Open markets analyze the freedom of trade and investment, customs duties, and restrictions on foreign investment. Together, these four categories offer crucial insights into how economic freedom influences national welfare and development (HERITAGE 2023). This section provides detailed information about the Economic Freedom Index, while the following section presents the practical aspect of the study.

6 Application

The study applies various machine learning methods to an imbalanced dataset, addressing a common challenge in predictive modeling. In this section, the presence of class imbalance is first illustrated using confusion matrices, which provide a clear depiction of misclassifications across different classes and the overall model performance. Given the impact of data imbalance on model accuracy, the Synthetic Minority Over-sampling Technique (SMOTE) was then employed to create a balanced dataset. The same machine learning methods were subsequently reapplied, enabling a direct comparison of results before and after implementing SMOTE. The confusion matrices from these applications revealed a significant reduction in the class imbalance problem, demonstrating SMOTE's effectiveness in enhancing model performance and yielding more reliable prediction outcomes.

Initially, the XGBoost method was used, resulting in a classification accuracy of 53% for the five-class problem. This low accuracy is attributed to the imbalanced data distribution. When analyzing the confusion matrix, it becomes evident that there is an irregular distribution of data, with certain classes being significantly underrepresented compared to others in the original dataset. This imbalance occurs because some examples from these minority classes are either sparsely represented or absent in the training dataset, which is constructed through random sampling from the original data. This lack of representation reduces the method's overall success and leads to increased misclassifications when the model is tested. To correct these irregularities, the SMOTE method was employed to balance the dataset, after which the method was reapplied. Following this adjustment, the classification accuracy improved to 79%. Table 1 shows the confusion matrix before and after SMOTE.

Next, the study applied the Support Vector Machines (SVM) method for classification. Initially, SVM was used to classify the imbalanced dataset, yielding an accuracy rate of 59%. When examining the confusion matrix, it is apparent that there is an irregular data distribution. To address this issue, the SMOTE method was used to balance the dataset, after which the Support Vector Machines (SVM) method was reapplied. On the balanced dataset, SVM achieved an improved accuracy rate of 69%. Table 2 shows the confusion matrix before and after SMOTE.

Table 1 XGBoost confusion matrix and SMOTE—XGBoost confusion matrix

	XGBoost confusion matrix					SMOTE—XGBoost confusion matrix				
	PL-0	PL-1	PL-2	PL-3	PL-4	PL- 0	PL-1	PL-2	PL-3	PL-4
TL-0	13	0	1	0	0	10	1	0	0	0
TL-1	1	0	2	0	1	0	6	2	0	0
TL-2	0	0	1	1	0	0	1	9	1	0
TL-3	0	0	3	2	3	1	0	2	5	4
TL-4	0	0	3	2	3	0	0	0	0	16

TL—true label, PL—predicted label

Table 2 SVM confusion matrix and SMOTE—SVM confusion matrix

	SVM confusion matrix					SMOTE—SVM confusion matrix				
	PL-0	PL-1	PL-2	PL-3	PL-4	PL-0	PL-1	PL-2	PL-3	PL-4
TL-0	*15*	3	1	0	0	*13*	4	1	0	0
TL-1	0	*1*	3	1	0	1	*11*	2	0	0
TL-2	2	0	*4*	0	0	0	2	*11*	1	1
TL-3	0	0	4	*5*	2	0	1	6	*9*	3
TL-4	0	0	3	3	*7*	0	0	0	5	*16*

TL—true label, PL—predicted label

Table 3 Accuracy rates of methods before and after SMOTE application

BEFORE SMOTE		AFTER SMOTE	
XGBOOST	SVM	XGBOOST	SVM
%53	%59	%79	%69

The results of the applications conducted using both methods indicated that more accurate classifications were achieved when the dataset had a balanced distribution. Specifically, the XGBoost method reached an accuracy rate of 79%, while the Support Vector Machines (SVM) method attained an accuracy rate of 69%. It was determined that XGBoost outperformed SVM when working with the balanced dataset. This finding aligns with previous studies in the literature (Ahsan et al. 2018; He et al. 2021). Table 3 presents the accuracy rates of both methods before and after applying the SMOTE technique.

According to the results presented in Table 3, the XGBoost method achieved more successful outcomes compared to the Support Vector Machines (SVM) method. After applying the SMOTE technique, the accuracy percentage of XGBoost increased more significantly. This finding suggests that XGBoost is more effective at learning from imbalanced datasets and yields better performance when class distribution is improved. Consequently, it can be concluded that XGBoost, when combined with SMOTE, offers a considerable advantage in classification performance.

In this section, the applications of XGBoost and SVM were conducted using data from the Economic Freedom Index. Initially, both methods were applied to the imbalanced dataset. Subsequently, the dataset was balanced using the SMOTE method, and the analyses were repeated. The findings from this application are discussed in detail in the next section, the conclusion, where recommendations are also provided.

7 Conclusion

In classification problems, especially when the number of classes is high, dealing with imbalanced data is a common challenge. This issue occurs when one class has significantly fewer instances than others, negatively impacting the performance of classification methods. To address this, various techniques have been developed, with the Synthetic Minority Over-sampling Technique (SMOTE) being one of the most prominent. SMOTE works by generating synthetic examples to create balanced datasets, thereby reducing class imbalance. It ensures that the synthetic examples retain the essential characteristics of the underrepresented classes. By increasing the number of instances for these minority classes, classification methods can achieve more accurate and reliable results.

In the application section of the study, the XGBoost and Support Vector Machines (SVM) classification methods were compared using data from the Financial Freedom Index 2022. The countries in the index were divided into five classes based on their rankings, and the classification methods were applied. Initially, both methods were used on the imbalanced dataset, but due to the uneven data distribution, they achieved low classification accuracy. To enable a fair comparison, the Synthetic Minority Over-Sampling Technique (SMOTE) was used to generate synthetic examples for the minority class, resulting in a balanced dataset. The XGBoost and SVM methods were then reapplied to this balanced dataset, which led to improved accuracy rates: 79% for XGBoost and 69% for SVM. These results indicate that XGBoost outperforms SVM in terms of classification performance, aligning with findings from similar studies in the literature (Ahsan et al. 2018; He et al. 2021; Wang et al. 2020).

Furthermore, the increase in accuracy was more pronounced for XGBoost after applying SMOTE, demonstrating that this method benefits significantly from class balancing techniques. The results underscore the effectiveness of using methods like SMOTE to address class imbalance and enhance the accuracy of machine learning models, particularly for powerful algorithms like XGBoost.

Based on these findings, several recommendations for future research on imbalanced data problems can be made. First, exploring various SMOTE variations, such as Borderline-SMOTE and SVM-SMOTE, which have shown success in imbalanced datasets, could improve model performance. Therefore, future studies could experiment with different SMOTE techniques. Additionally, alternative sampling methods beyond SMOTE, such as Random Oversampling, Random Undersampling, and ADASYN, should be examined to understand their effects on model accuracy. Comparing these techniques could help identify the most suitable approach for specific dataset characteristics.

Another important suggestion is to perform hyperparameter tuning on the methods. Conducting studies on hyperparameter optimization, using techniques like Grid Search or Random Search, can enhance the performance of XGBoost and SVM. Lastly, the study recommends evaluating other machine learning models, such as LightGBM, Random Forest, and CatBoost, which may perform well on imbalanced datasets, for future research. To increase the generalizability and validate the methods used in this study, it would be beneficial to test these methods on different imbalanced datasets as well. Implementing these recommendations can contribute to developing effective solutions for imbalanced data problems and achieving higher accuracy rates.

References

Abdeltawab H, Khalifa F, ElNakieb Y, Elnakib A, Taher F, Alghamdi NS, Sandhu HS, El-Baz A (2022) Predicting the level of respiratory support in COVID-19 patients using machine learning. Bioengineering 9(10):536. https://doi.org/10.3390/bioengineering9100536

Agrawal A, Viktor HL, Paquet E (2015) SCUT: multi-class imbalanced data classification using SMOTE and cluster-based undersampling. In: 2015 7th International joint conference on knowledge discovery, knowledge engineering and knowledge management (IC3k), pp 226–234

Ahsan M, Gomes R, Denton A (2018) SMOTE implementation on phishing data to enhance cybersecurity. In: 2018 IEEE International conference on electro/information technology (EIT), 0531–0536. https://doi.org/10.1109/EIT.2018.8500086

Akutsu T, Hayashida M, Tamura T (2012) Finding conserved regions in protein structures using support vector machines and structure alignment. In: Shibuya T, Kashima H, Sese J, Ahmad S (Eds) Pattern recognition in bioinformatics, vol 7632. Springer, Berlin Heidelberg, pp 233–242. https://doi.org/10.1007/978-3-642-34123-6_21

Atsauri MR, Mawengkang H, Efendi S (2023) Enhancing unbalanced data classification with cross-validation and extreme gradient boosting: a comprehensive analysis. J Inform Telecommun Eng 7(1):30–42. https://doi.org/10.31289/jite.v7i1.8690

Belanová K, Sivák R, Dziura B, Bastyr S, Šípová M (2023) Empirical analysis of the impact of economic freedom on economic growth in the Slovak Republic, the Czech Republic and Singapore. Entrepreneurship Sustain Issues 11(2):419–432. https://doi.org/10.9770/jesi.2023.11.2(28)

Ben-Hur A, Ong CS, Sonnenburg S, Schölkopf B, Rätsch G (2008) Support vector machines and kernels for computational biology. PLoS Comput Biol 4(10):e1000173. https://doi.org/10.1371/journal.pcbi.1000173

Binson VA, Subramoniam M, Sunny Y, Mathew L (2021) Prediction of pulmonary diseases with electronic nose using SVM and XGBoost. IEEE Sens J 21(18):20886–20895. https://doi.org/10.1109/JSEN.2021.3100390

Blagus R, Lusa L (2013) SMOTE for high-dimensional class-imbalanced data. BMC Bioinform 14(1):106. https://doi.org/10.1186/1471-2105-14-106

Brkić I, Gradojević N, Ignjatijević S (2020) The impact of economic freedom on economic growth? New European dynamic panel evidence. J Risk Financ Manag 13(2):26. https://doi.org/10.3390/jrfm13020026

Cervantes J, Garcia-Lamont F, Rodríguez-Mazahua L, Lopez A (2020) A comprehensive survey on support vector machine classification: applications, challenges and trends. Neurocomputing 408:189–215. https://doi.org/10.1016/j.neucom.2019.10.118

Chawla NV, Bowyer KW, Hall LO, Kegelmeyer WP (2002) SMOTE: synthetic minority over-sampling technique. J Artif Intell Res 16:321–357. https://doi.org/10.1613/jair.953

Chen T, Guestrin C (2016) XGBoost: a scalable tree boosting system. In: Proceedings of the 22nd ACM SIGKDD international conference on knowledge discovery and data mining, pp 785–794. https://doi.org/10.1145/2939672.2939785

Cui S, Youn E, Lee J, Maas SJ (2014) An improved systematic approach to predicting transcription factor target genes using support vector machine. PLoS ONE 9(4):e94519. https://doi.org/10.1371/journal.pone.0094519

Czarnowski I (2022) Weighted ensemble with one-class classification and over-sampling and instance selection (WECOI): an approach for learning from imbalanced data streams. J Comput Sci 61:101614. https://doi.org/10.1016/j.jocs.2022.101614

Erilli NA (2018) Economic freedom index calculation using FCM. Alphanumeric J 6(1):93–116. https://doi.org/10.17093/alphanumeric.337322

Evans A, Naurodski S (2019) Ranking Belarus on competitiveness and economic freedom. J Entrepreneurship Public Policy 8(2):226–240. https://doi.org/10.1108/JEPP-07-2019-108

Fernández A, García S, Galar M, Prati RC, Krawczyk B, Herrera F (2018) Learning from imbalanced data sets. Springer International Publishing. https://doi.org/10.1007/978-3-319-98074-4

Gao Q, Jin X, Xia E, Wu X, Gu L, Yan H, Xia Y, Li S (2020) Identification of orphan genes in unbalanced datasets based on ensemble learning. Front Genet 11:820. https://doi.org/10.3389/fgene.2020.00820

Gu K, Wang J, Qian H, Su X (2021) Study on intelligent diagnosis of rotor fault causes with the PSO-XGBoost algorithm. Math Probl Eng 2021:1–17. https://doi.org/10.1155/2021/9963146

Gu Q, Tian J, Li X, Jiang S (2022) A novel Random Forest integrated model for imbalanced data classification problem. Knowl-Based Syst 250:109050. https://doi.org/10.1016/j.knosys.2022.109050

Gwartney J, Lawson R, Hall J, Murphy R (2022) Economic freedom of the world: 2022 annual report (2022nd ed). Fraser Institute. https://doi.org/10.53095/88975001

Hasib KMd, Rahman F, Hasnat R, Alam MdGR (2022) A machine learning and explainable AI approach for predicting secondary school student performance. In: 2022 IEEE 12th Annual computing and communication workshop and conference (CCWC), pp 0399–0405. https://doi.org/10.1109/CCWC54503.2022.9720806

He J, Hao Y, Wang X (2021) An interpretable aid decision-making model for flag state control ship detention based on SMOTE and XGBoost. J Mar Sci Eng 9(2):156. https://doi.org/10.3390/jmse9020156

HERITAGE (2023) Index of economic freedom. https://www.heritage.org/index/about

Jiang Y, Wang H, Xie Q (2009) Classification model of companies' financial performance based on integrated support vector machine. In: 2009 International conference on management science and engineering, pp 1322–1328. https://doi.org/10.1109/ICMSE.2009.5318030

Li L (2015) Economic freedom and foreign direct investment in BRICS countries. In: Third international conference on advances in social science, Management and human behavior—SMHB 2015, pp 61–64. https://doi.org/10.15224/978-1-63248-067-5-73

Lin W, Tian X, Lu X, Ma D, Wu Y, Hong J, Yan R, Feng G, Cheng Z (2022) Prediction of bedridden duration of hospitalized patients by machine learning based on EMRs at admission. CIN: Comput, Inform, Nurs 40(4):251–257. https://doi.org/10.1097/CIN.0000000000000765

Mateo F, Tarazona A, Mateo EM (2021) Comparative study of several machine learning algorithms for classification of unifloral honeys. Foods 10(7):1543. https://doi.org/10.3390/foods10071543

Morde V (2019) XGBoost algorithm: Long May She Reign! | by Vishal Morde | Towards Data Science. https://towardsdatascience.com-medium-com-vishalmorde-xgboost-algorithm-long-she-may-rein-edd9f99be63d

Naanwaab C (2013) Determinants of productivity in Africa: the role of economic freedom. J Dev Agric Econ 5(10):390–402. https://doi.org/10.5897/JDAE12.162

Neelamegam S, Ramaraj E (2013) Classification algorithm in data mining: an overview. Int J P2P Netw Trends Technol (IJPTT) 4(8):369–374

Njoroge E, Alty SR, Gani MR, Alkatib M (2006) Classification of cervical cancer cells using FTIR Data. In: 2006 International conference of the IEEE engineering in medicine and biology society, pp 5338–5341. https://doi.org/10.1109/IEMBS.2006.260024

Ott J (2018) Measuring economic freedom: better without size of government. Soc Indic Res 135(2):479–498. https://doi.org/10.1007/s11205-016-1508-x

Pertea M, Mount SM, Salzberg SL (2007) A computational survey of candidate exonic splicing enhancer motifs in the model plant Arabidopsis thaliana. BMC Bioinform 8(1):159. https://doi.org/10.1186/1471-2105-8-159

Pisner DA, Schnyer DM (2020) Support vector machine. In Machine learning. Elsevier, pp 101–121. https://doi.org/10.1016/B978-0-12-815739-8.00006-7

Shen Q, Shi W-M, Kong W, Ye B-X (2007) A combination of modified particle swarm optimization algorithm and support vector machine for gene selection and tumor classification. Talanta 71(4):1679–1683. https://doi.org/10.1016/j.talanta.2006.07.047

Torlay L, Perrone-Bertolotti M, Thomas E, Baciu M (2017) Machine learning–XGBoost analysis of language networks to classify patients with epilepsy. Brain Inform 4(3):159–169. https://doi.org/10.1007/s40708-017-0065-7

Vitenu-Sackey PA (2022) Economic freedom, inclusive growth, and financial development: a heterogeneous panel analysis of less financially developed countries. Review. https://doi.org/10.21203/rs.3.rs-1791868/v1

Wang S, Liu S, Zhang J, Che X, Yuan Y, Wang Z, Kong D (2020) A new method of diesel fuel brands identification: SMOTE oversampling combined with XGBoost ensemble learning. Fuel 282:118848. https://doi.org/10.1016/j.fuel.2020.118848

Xia J, Caragea D, Brown S (2008) Exploring Alternative Splicing Features Using Support Vector Machines. IEEE International Conference on Bioinformatics and Biomedicine 2008:231–238. https://doi.org/10.1109/BIBM.2008.12

Yadav S, Bhole GP (2020) Handling imbalanced dataset classification in machine learning. In: 2020 IEEE Pune Section international conference (PuneCon), pp 38–43. https://doi.org/10.1109/PuneCon50868.2020.9362471

Yevdokimov Y, Melnyk L, Lyulyov O, Panchenko O, Kubatko V (2018) Economic freedom and democracy: determinant factors in increasing macroeconomic stability. Prob Perspect Manag 16(2):279–290. https://doi.org/10.21511/ppm.16(2).2018.26

Yoo I, Alafaireet P, Marinov M, Pena-Hernandez K, Gopidi R, Chang J-F, Hua L (2012) Data mining in healthcare and biomedicine: a survey of the literature. J Med Syst 36(4):2431–2448. https://doi.org/10.1007/s10916-011-9710-5

Yusri HIH, Ab Rahim AA, Hassan SLM, Halim ISA, Abdullah NE (2022) Water quality classification using SVM And XGBoost method. In: 2022 IEEE 13th control and system graduate research colloquium (ICSGRC), pp 231–236. https://doi.org/10.1109/ICSGRC55096.2022.9845143

Zhang L, Zhan C (2017) Machine learning in rock facies classification: an application of XGBoost. In: International geophysical conference, Qingdao, China, 17–20 April 2017, pp 1371–1374. https://doi.org/10.1190/IGC2017-351

Zhao XS, Bao LL, Ning Q, Ji JC, Zhao XW (2018) An improved binary differential evolution algorithm for feature selection in molecular signatures. Mol Inf 37(4):1700081. https://doi.org/10.1002/minf.201700081

The Impact of ESG Factors on the Propensity for Dividends for European Firms: A Machine Learning Approach

Önder Dorak and Aykut Yakar

Abstract This study examines the influence of Environmental, Social, and Governance (ESG) factors on the propensity of European firms to pay dividends, using a machine learning approach for classification. Employing data from 1886 publicly traded firms in 2023, the analysis identifies the most influential ESG sub-dimensions and assesses their role in predicting dividend payouts. The findings highlight that environmental factors, particularly emissions, resource use, and innovation, have the strongest positive impact. Gradient boosting proved to be the most effective model, balancing predictive accuracy and interpretability. Using SHAP values, the study identifies key ESG sub-dimensions and their contributions to dividend payout predictions. This research advances the understanding of ESG factors' role in dividend policy by focusing on sub-dimensions and integrating machine learning for enhanced predictive and explanatory insights.

Keywords ESG factors · Dividend policy · Machine learning · Feature selection · Regularization

1 Introduction

Dividend policy includes decisions on how much of the income generated by a firm will be retained within the firm and how much will be distributed to shareholders. The decision to pay dividends or retain earnings within the firm for future investments is mainly expected to impact the firm's projected cash flows and cost of capital. Directing the retained earnings toward profitable investment projects will increase the firm's projected cash flows. On the other hand, retained earnings will inevitably

Ö. Dorak (✉)
The Faculty of Economics and Administrative Sciences, Anadolu University, Eskişehir, Türkiye
e-mail: odorak@anadolu.edu.tr

A. Yakar
The Faculty of Economics and Administrative Sciences, Anadolu University, Eskişehir, Türkiye
e-mail: aykut_yakar@anadolu.edu.tr

M. Gün and B. Kartal (eds.), *Machine Learning in Finance*, Contributions to Finance and Accounting, https://doi.org/10.1007/978-3-031-83266-6_7

have an impact on shareholders' expectations. In dividend decisions, the profitability of the future investments and investors' expectations regarding these investments are expected to influence investors' expected returns and, consequently, the firm's cost of capital. Similarly, when the firm distributes dividends, it may be perceived as less risky by investors, which will likewise impact the cost of capital. However, when a firm with profitable investment opportunities distributes dividends, it will create a need for external financing. Since dividend decisions represent an issue of conflict between the interests of the firm and its investors, it is a critical decision area where financial managers must make an optimal choice between distributing dividends and retaining profits within the firm.

In determining the dividend policy, the firm's investment opportunities and investor expectations stand out as key factors. However, the factors influencing dividend policy are not limited to these two elements alone. Factors such as profitability, cash flow situation, liquidity, debt structure, economic conditions, as well as legal and tax obligations, also play a significant role in these decisions. The effects of information in non-financial reports, such as ESG have been increasingly studied in the literature regarding dividend policy in recent times. The growing concerns of stakeholders regarding sustainability have led to an increased interest in reports like ESG when making decisions about firms.

ESG reporting is most simply defined as the public disclosure of a firm's environmental, social, and governance practices. It provides extensive information on the firm's approach to environmental, social, and governance issues (Kartal et al. 2024). Disclosing information about practices that impact investors, society, employees, the environment, and customers has made ESG reports crucial in meeting investors' demand for non-financial information (Ellili 2022). Thus, it has become increasingly important to provide stakeholders with more comprehensive information on ESG-related matters, as it is no longer sufficient to inform them solely with reports containing financial information (Biondi and Bracci 2018).

ESG reporting's environmental (E) component encompasses the firm's efforts to create positive impacts on the environment, through compliance with existing regulations and recognition of future impacts. Social (S) activities involve fair treatment of primary stakeholders and the safeguarding of the social environment in which the firm operates. Governance (G) encompasses corporate ethics and integrity, focusing on principles like transparency, fairness, and the efficient operation of the board of directors (Limkriangkrai et al. 2017).

In recent years, the interest of all stakeholders in firms' ESG performance has grown rapidly. This increased interest has placed pressure on firms to improve their sustainability practices (Ellili 2022). As a result, firms have begun integrating ESG criteria into various aspects of their business to enhance their reputation and gain a competitive advantage (Bilyay Erdogan et al. 2023). Thus, firms' ESG disclosures and actions have shifted from being merely ethical considerations to becoming critical factors that support their long-term existence and competitiveness. Similarly, investors and stakeholders now view ESG integration not merely as a moral need but as a strategic imperative with the ability to enhance a firm's risk management, operational efficiency, and financial performance (Ananzeh et al. 2024).

2 The General Framework of Dividend Theories and ESG

In traditional finance theory, the primary goal of firms is to maximize investors' wealth through dividend distribution and capital gains. However, the increasing importance attributed to ESG has created a need for firms to consider the interests of a broader range of stakeholders, such as customers, communities, employees, and the environment (Ananzeh et al. 2024). This shift is supported by Freeman's (1984) stakeholder theory, which argues that firms should meet the expectations of all stakeholders, not only shareholders. Therefore, both ESG initiatives and dividend distributions play a vital role in a firm's success, serving as channels to signal its quality and enhance its reputation (Bilyay Erdogan et al. 2023).

The relationship between dividend policy and ESG performance can be explained by theories based on asymmetric information, such as the signaling theory by Bhattacharya (1979) and agency theory by Easterbrook (1984) and Jensen (1986). According to signaling theory, dividend payments that exceed expectations are perceived as a positive signal regarding the firm's current and future earnings. If the ESG ratings and dividend payments act as complementary signals, improvement in the ESG performance, similar to dividend payments, will also be perceived as a positive signal in the market (Maquieira et al. 2024). In cases where the two signals are complementary, dividend distribution serves as a signal of the firm's financial strength in the market, while improvements in ESG ratings help convince investors of the firm's sustainability in terms of environmental, social, and governance aspects.

Dividend decisions and ESG performance can serve as complementary signals; however, due to various reasons, primarily financial constraints, they can also act as substitute signals. According to signaling theory, a negative relationship between dividend policy and ESG performance arises when dividend policy and ESG performance are perceived as substitute signals (Maquieira et al. 2024). Since the firm has decided to distribute high dividends, it will have to limit its ESG investments due to financial constraints. In such a case, the firm will likely attract investors who expect immediate returns and are less concerned with the firm's long-term ESG performance. In the opposite scenario, the firm could increase its ESG investments by distributing fewer or no dividends. In this case, the firm's dividend and ESG investment decisions would be positively received by investors who prioritize sustainability. In both scenarios, dividend policy and ESG performance act as substitute signals in line with the firm's priorities and the expectations of different investor groups, resulting in a negative relationship between them.

According to agency theory, dividend distribution is seen as a crucial tool for preventing agency costs that arise from managing the firm's free cash flows and reducing the inefficient use of excess cash by managers. While ESG investments are highly valuable for the firm's sustainability, they can also be a source of agency costs. Managers may make ESG investments beyond the firm's actual needs to enhance their image of responsibility toward ESG, driven by their interests (Ananzeh et al. 2024). In such a case, excess cash flows directed toward ESG investments would result in lower dividend payments, leading to agency costs between managers and shareholders (Maquieira et al. 2024). However, if managers undertake ESG investments in alignment with all stakeholders' expectations and the firm's objectives, potential agency costs can be avoided. This approach enables building more trustworthy relationships which leads to improved performance and long-term profitability, ultimately raising the firm ability to distribute more cash dividends (Ananzeh et al. 2024). Agency theory suggests that, when the interests of managers and stakeholders are balanced through dividend policy and ESG investments, the relationship between them will be positive in the long term.

This study has two main objectives. Firstly, it aims to reveal which ESG pillars and sub-dimensions have the highest impact on the propensity to pay dividends. The second objective is to decide which supervised machine learning algorithm for classification outperforms others when analyzing the relationship between ESG categories and the dividend payout. To achieve this goal, this chapter uses different machine learning methods to classify dividend-paying and non-dividend-paying firms based on ESG categories with a sample of 1886 publicly traded European firms. Due to Europe's leading role in significant regulatory developments regarding environmental, social, and governance issues, the chapter analyzes firms publicly traded in Europe.

This study is expected to contribute to the literature in two ways. Previous research has generally focused on the relationship between ESG performance and dividend payments, while the firm's propensity to pay dividends has received less attention in the literature. Moreover, to the authors' knowledge, these studies have only utilized ESG pillars, and the impact of their sub-dimensions on the propensity to pay dividends has not been explored. This study is expected to fill the gap in the literature by providing a better understanding of the influence of ESG sub-dimensions on the propensity to pay dividends. Finally, to the authors' knowledge, no study in the literature examining ESG performance and dividend payout propensity has employed machine learning methods. In this regard, this study is considered the first to analyze the relationship between ESG performance and the propensity to pay dividends by benefiting from the advantages offered by machine learning methods.

3 Literature Review

Many studies have examined the effects of ESG integration on financial performance, firm value, investment efficiency, profitability, and cost of capital. The relationship between ESG performance and financial performance has been one of the most frequently examined topics in ESG-related literature. In the literature, interest in the relationship between dividend payouts and ESG performance has increased in recent years. However, there is no clear consensus in the literature regarding the relationship between dividend payouts and ESG performance. Nevertheless, the number of studies in the literature reporting that ESG performance positively affects dividend payouts is greater than those reporting negative effects.

Firms should act in a way that considers not only the interests of investors but also those of all stakeholders while conducting their operations and investments. Accompanying ESG investments made on this basis with high dividend payouts demonstrates that the firm considers all its stakeholders. Thus, high ESG performance and high dividends will be perceived by the market as complementary and a positive signal according to the signaling theory. Hussainey and Walker (2009), Matos et al. (2020), Trihermanto and Yunieta (2020), Bilyay Erdogan (2023), and Dahiya et al. (2023) reported a positive relationship between dividend payouts and ESG performance, as firms provide more information to their investors through ESG reports.

ESG initiatives create value for shareholders by building trust and fostering collaboration with stakeholders, which in turn strengthens their support for the firm's operations. A firm's efforts to align its ESG investments with stakeholder expectations and at the best possible scale can help reduce information asymmetry, as well as the agency costs associated with the firm's free cash flows. Since higher dividend distributions also contribute to reducing agency costs, ESG investments made in line with stakeholder expectations are anticipated to have a positive relationship with dividends, according to agency theory (Ananzeh et al. 2024; Maquieira et al. 2024). Benlemlih (2019), Hasan and Habib (2020), Dimitropoulos and Koronios (2021), and Ellili (2022) concluded that high dividend payments reduce agency costs, while ESG investments are not seen as a waste of resources and effectively meet stakeholder expectations.

According to signaling theory, investors will demand higher dividends from firms that provide insufficient information disclosure due to the resulting information asymmetry. Thus, firms may choose to make more ESG investments to address information asymmetry and agency problems, which could lead to lower dividend payouts or vice versa (Ellili 2022). The studies by Ni and Zhang (2019) and Saeed and Zamir (2021) report, from the perspective of signaling theory, that dividends and ESG act as substitute signals and that there is a negative relationship between them.

Studies on ESG performance and dividend distribution have generally examined the impact of ESG performance on dividends, the stability of dividends, the moderating role of various drivers in the link between ESG scores and dividend policy, and the propensity to pay dividends. Similar to this study, Cheung et al. (2018), Matos et al. (2020), Sheikh et al. (2021), Matuszewska-Pierzynka et al. (2023) and Shear et al. (2024)'s works also sought to explain firms' propensity to pay dividend through their ESG performance. Sheikh et al. (2021) and Shear et al. (2024) reported that firms with high ESG scores are more propensity to pay dividends, whereas Matuszewska-Pierzynka et al. (2023) obtained the contradictory result. On the other hand, Cheung et al. (2018) and Matos et al. (2020) reported no relationship between ESG scores and the propensity to pay dividends.

Some studies examining the relationship between dividend payout and ESG performance have included not only ESG scores but also the pillars covered in ESG reports in their analysis. Matos et al. (2020), Zadeh (2021), Zahid et al. (2023), Bilyay-Erdogan et al. (2023), Ananzeh et al. (2024), Shear et al. (2024) and Salvi et al. (2024) reported a positive relationship between the environmental dimension and dividend payout, while Matuszewska-Pierzynka et al. (2023) reported a negative effect. Zadeh (2021), Salah and Amar (2022), Zahid et al. (2023), Bilyay-Erdogan et al. (2023), Ananzeh et al. (2024) and Salvi et al. (2024) found that high scores in the social dimension were associated with increased dividend payouts, whereas Matuszewska-Pierzynka et al. (2023) obtained the opposite results. The impact of governance on dividend payout was reported as positive in studies by Matos et al. (2020), Salah and Amar (2022), Zahid et al. (2023), Bilyay-Erdogan et al. (2023) and Matuszewska-Pierzynka et al. (2023) and Ananzeh et al. (2024) although there are studies where the effect was reported as negative.

In the literature, only one study has been identified that examines the dimensions of ESG pillars in the context of the relationship between dividend distribution and ESG performance. Lakhal et al. (2023) used observations from 60 countries in their study covering the years 2003–2009. The study reported that dividend payments were positively associated with all ESG dimensions except for the shareholders' dimension. Lakhal et al. (2023) also examined the moderating effect of the shareholder dimension on the relationship between dividend payout and ESG performance. It was concluded that the shareholder dimension has a negative moderating effect on the relationship between dividend payout and environmental pillar and emissions and CSR strategy sub-dimensions.

4 Analysis of the Impacts of ESG Factors on Dividend Policies of European Firms

This section firstly explains the data and the variables and later presents the machine learning methodologies used in the analysis. Afterwards, findings about the impacts of ESG factors on dividend policies of European firms are explained.

4.1 Data

The data in this study consists of 1886 publicly traded European firms from eight industries (excluding finance and real estate firms) and 43 stock exchange markets for the year 2023. The data consists of 1219 dividend-paying firms and 667 non-paying dividend firms. So, 64.6% of European firms in the selected sectors paid dividends to their shareholders in the year 2023. In the dataset, there are 1886 firms across eight sectors including basic materials, consumer cyclicals, energy, healthcare, industrials, noncyclicals, technology, and utilities.

The data is retrieved from the LSEG Eikon (London Stock Exchange, previously Refinitiv Eikon) database. LSEG Eikon provides a comprehensive ESG database that has extensive use in academia. According to Berg et al. (2021), ESG data of LSEG have been used in more than 1500 academic research since 2003. LSEG uses various resources including annual reports, company websites, CSR reports, news sources, etc. when they are calculating their ESG scores (LSEG Data and Analytics 2023). LSEG uses over 630 ESG metrics in their ESG calculation process and they are grouped into 10 categories under three pillars with different weightings: Environmental, Social, and Governance. The environmental pillar includes metrics for resource use (20%), emissions (28%), and innovation (20%). The social pillar covers the workforce (30%), human rights (8%), community (14%), and product responsibility (10%). Finally, the governance pillar includes management (35%), shareholders (12%), and CSR strategy (9%) (LSEG Data and Analytics 2023).

LSEG uses percentile rank scoring methodology which means every firm has a category score based on its relative performance to other firms in the same sector. It also means that every ESG score ranges between 0 and 100. As a result of the calculation process, every firm receives an ESG combined score, three pillar scores, ten category scores, and one controversy score.

This study aims to investigate the impacts of ESG scores on dividend payout policies. To achieve this purpose, the dividend payout, a binary variable, was chosen as the target variable. The target variable was then modeled with the help of 10 ESG categories under the three ESG pillars and the ESG controversy score through various classification algorithms.

4.2 Methodology

One of the purposes of this study is to assess the effectiveness of various supervised machine learning algorithms in classifying firms based on their likelihood of paying dividends, using their ESG (Environmental, Social, and Governance) scores as predictors. Specifically, this study aims to determine how accurately these models can distinguish between firms that distribute dividends and those that do not, based solely on their ESG performance indicators. By applying a range of classification techniques, the authors seek to identify the most suitable algorithms for this task, while also examining the relative importance of different ESG factors in influencing dividend payout behavior.

This study exploits several classification methods used in supervised machine learning. Classification methods aim to assign samples to predefined categories based on the characteristics of the predictor variables. Different approaches achieve this objective by employing various algorithmic and mathematical techniques. Some methods, like logistic regression, produce linear and easy-to-interpret models, allowing decision-makers to understand the relationships between predictors and the target variable clearly. However, these models are not able to identify the relationships where the dataset is complex or non-linear (Kuhn and Johnson 2013).

In contrast, more advanced techniques like support vector machines (SVM), k-nearest neighbors (k-NN), and neural networks excel at capturing non-linear interactions between features, often resulting in higher predictive accuracy. Despite their strong performance, these models are less interpretable because they do not explain how the predictions are made, earning them the label of "black-box" models (Li et al. 2022). On the other hand, tree-based models, such as decision trees, strike a balance by providing a degree of interpretability while being capable of modeling non-linear relationships. Decision trees visually display the decision-making process, making it easier to understand how the features contribute to the final classification. However, decision trees are prone to high variance, changing one observation in a data set may change the result altogether. Therefore, a single decision tree is not generally preferred for generalization (Hastie et al. 2001).

Ensemble methods such as bagging, gradient boosting and random forests have been developed to reduce bias and variance, increasing prediction accuracy (Russell and Norvig 2016). They also address the overfitting problem inherent in a single decision tree. These methods combine multiple trees to form a stronger, more robust model. By aggregating the predictions of many decision trees, ensemble models can significantly improve accuracy while reducing overfitting, offering a more stable and generalizable approach to classification. In addition, techniques like bagging and boosting further enhance the performance of tree-based models, mitigating their instability and enhancing their predictive power, while offering some level of interpretability compared to black-box models (Hastie et al. 2001).

In summary, while linear models provide simplicity and transparency, they may lack flexibility and accuracy. Non-linear models, on the other hand, offer superior predictive performance but are often less interpretable. Ensemble methods provide a compelling middle ground, offering both strong predictive performance and some interpretability, making them well-suited for complex classification tasks.

4.3 Results

The supervised machine learning procedure includes several consecutive steps. These are splitting the data into training, and testing sets; hyperparameter tuning, estimating the final model by using hyperparameters selected on training data, and estimating prediction performance on test data (Raschka 2020).

As a first step, 70% of the data is split for training and validation, and 30% is split for the testing procedure. In the second step, tenfold cross-validation is applied on the training set for different hyperparameter configurations to select the best hyperparameters. Table 1 presents the best hyperparameters selected for each classification algorithm used in the study. Hyperparameters show that the models are typically more conservative, with higher regularization and lower depths (e.g., high number of neighbors in k-NN, few hidden units for neural networks).

Based on the best hyperparameters in Table 1, the final models are estimated on full training data. Afterward, the estimated final models are tested on the test data. Tables 2 and 3 present ESG category score results under various classification metrics i.e. accuracy, precision, recall, f-score, and ROC_AUC (area under the curve).

The model comparison in Tables 2 and 3 reveals that in modeling the dividend payout based on ESG categories, most of the classification models demonstrate

Table 1 Best hyperparameters of classification algorithms

Model	Best hyperparameters
Random forest	$mtry(m) = 1$
Gradient boosting	$treedepth(d) = 2 learnrate(\eta) = 0.008 lossreduction(\gamma)$=0.091
Support vector machine	$cost(C) = 0.0049$
Neural network	$hiddenunits(h) = 8$ $penalty(\lambda) = 0.949$ $epochs(e) = 83$
Naive Bayes	$smoothness(s) = 0.571$ $Laplace(L) = 2.86$
Regularized logistic regression	$penalty(\lambda) = 0.0002$ $mixture(\alpha) = 0.207$
K-nearest neighbors	$neighbors(k) = 15$
Logistic regression	None
Decision trees	$treedepth(d) = 5$ $costcomplexity(\alpha) = 1e - 5$

Table 2 Model performances of final models on training data

Model	Training					
	Accuracy	Specificity	Sensitivity	Precision	F-score	ROC-AUC
Random forest	0.9871	0.9812	0.9979	0.9988	0.9899	0.9978
Gradient boosting	0.7400	0.9472	0.3605	0.7306	0.8249	0.8031
Support vector machines	0.7013	0.9484	0.2489	0.6980	0.8042	0.7119
Neural networks	0.7218	0.9426	0.3176	0.7166	0.8142	0.7297
Naive Bayes	0.6998	0.7843	0.5451	0.7594	0.7716	0.7261
Regularized logistic regression	0.6998	0.8652	0.3970	0.7242	0.7885	0.7163
K-nearest neighbors	0.7741	0.9109	0.5236	0.7778	0.8391	0.8747
Logistic regression	0.6990	0.8652	0.3948	0.7235	0.7880	0.7160
Decision tree	0.7286	0.9461	0.3305	0.7212	0.8185	0.7410

stable results across both testing and training tests. One specific exception is the random forest model having extremely high accuracy (0.9871) on training data and significantly lower accuracy (0.7284) on test data indicating overfitting. In the other methods, the training and testing accuracies are not so diverged indicating a good generalization ability.

Logistic regression and regularized logistic regression also offer balanced performance, demonstrating stable results across both training and testing datasets. These two methods provide a good trade-off between sensitivity and specificity measures. They have the highest accuracy and AUC metrics with gradient boosting. However, their accuracy on the testing data improved (similar to support vector machines), in contrast to the expectations, which possibly stemmed from some of the observations in the training set are hard to learn for these methods. Also, logistic regression

Table 3 Model performances of final models on test data

Model	Testing					
	Accuracy	Specificity	Sensitivity	Precision	F-score	ROC-AUC
Random forest	0.7284	0.9180	0.3831	0.7304	0.8136	0.7410
Gradient boosting	0.7231	0.9262	0.3532	0.7228	0.8120	0.7335
Support vector machines	0.7284	0.9672	0.2935	0.7137	0.8213	0.7264
Neural networks	0.7160	0.9426	0.3035	0.7113	0.8108	0.7251
Naive Bayes	0.6966	0.7978	0.5124	0.7487	0.7725	0.7301
Regularized logistic regression	0.7178	0.8880	0.4080	0.7320	0.8025	0.7331
K-nearest neighbors	0.6984	0.8661	0.3930	0.7221	0.7876	0.6918
Logistic regression	0.7178	0.8880	0.4080	0.7320	0.8025	0.7323
Decision tree	0.7090	0.9290	0.3085	0.7098	0.8047	0.6982

and regularized logistic regressions almost have identical metric results that the regularization process did not significantly contribute in prediction accuracy.

k-NN performs well on the training set but shows a drop in test accuracy, indicating slight overfitting. It has reasonable sensitivity and specificity, making it a balanced but not optimal choice. The decision tree also overfits the training data, as indicated by the drop in test performance, particularly in specificity. Its high sensitivity is a strength, but low specificity may limit its use in applications requiring balanced classification.

In general, ML algorithms showed balanced performance across training and test sets. Although the results vary across the models, they have high specificity, low sensitivity, and moderately high precision. High specificity suggests that the models can correctly identify non-paying dividend payers. On the other hand, low sensitivity underlies that ML models have a conservative approach, leading to incorrect predictions of dividend-paying firms. Moderately high precision implies that it is generally accurate when the model predicts a positive case (dividend-paying firm). This pattern suggests that the models are generally conservative in predicting positive cases (dividend payers), tending to classify instances as non-dividend payers unless there is strong evidence to the contrary. As a result, while the models effectively minimize false positives, they may overlook some true positive cases.

Given that, in the original dataset, 1219 observations (64.6%) are dividend-paying firms and 667 observations are non-dividend-paying firms (35.4%), the performance results in Table 3 can be compared with the base model (or naïve classifier). Most of the ML models achieved slightly higher accuracy and precision in training and test sets than the base model.

Gradient Boosting generalizes fairly well with comparable performance on training and test sets, it has the highest balance between accuracy, sensitivity, and ROC AUC across training and testing sets, showing minimal overfitting. Similar performances in training and testing data sets mean the gradient boosting model generalizes well and achieves underlying patterns in the data. Also, its training ROC AUC of 0.8031 and testing ROC AUC of 0.7335 shows it reliably distinguishes between dividend and non-dividend payer firms. Overall, Gradient Boosting offers a reliable combination of generalizability, balanced metrics, and robustness, making it the optimal choice for this study.

4.4 Interpretability

In this subsection, the interpretability of the gradient boosting model found in the previous section is investigated further to classify firms whether they pay dividends or not based on ESG category scores. Much of machine learning studies focus on the prediction accuracy of their models. Yet, complex models such as ensemble models or neural networks lack of interpretability of the predictors. The ones who provide interpretability, on the other hand, such as logistic regression overlook the nonlinear dependencies. Also, models like decision trees providing great interpretability for the decision-makers are heavily sensitive to every instance in the data set so they do not achieve robust generalizations. Although the prediction ability of a machine learning model has the utmost importance to judge whether the model has a good performance, in social studies such as finance interpretation of the predictors is equally important to gain insights and expand knowledge.

To overcome this tension between accuracy and interpretability in complex machine learning models, Lundberg and Lee (2017) presented the SHAP (SHapley Additive exPlanations) technique. The Shapley value was first coined by (Shapley 1953) as a method in coalition game theory, and then adapted to machine learning applications. The Shapley value in machine learning assesses the importance of each feature by quantifying how it affects the model's prediction. For a given set of feature values, the Shapley value estimates how much a particular feature contributes to the difference between the actual prediction and the average prediction across the dataset. The SHAP value is an additive function of the features considered. The sum of the contributions of features equals the difference between the actual prediction and the average prediction (the base value).

The Shapley values have been used in finance literature several times. Del Vitto et al. (2023) used the Shapley values when investigating the explainability of the ESG ratings in their black box machine learning models. Coulombe Goulet et al. (2024) used Shapley values to identify the most significant determinants of cross-sectional expected returns in their machine-learning-based portfolio performance study.

Figure 1 displays the feature importance which displays the overall assessment of feature importance. In the feature importance plot, the X-axis displays the mean SHAP values for each feature. A higher SHAP value indicates that the feature has a stronger impact on the predictions of the model. The plot shows that CSR strategy, emissions, and resource use are the top three features having the greatest impacts on model output. Innovation, shareholders, and controversy scores have a relatively

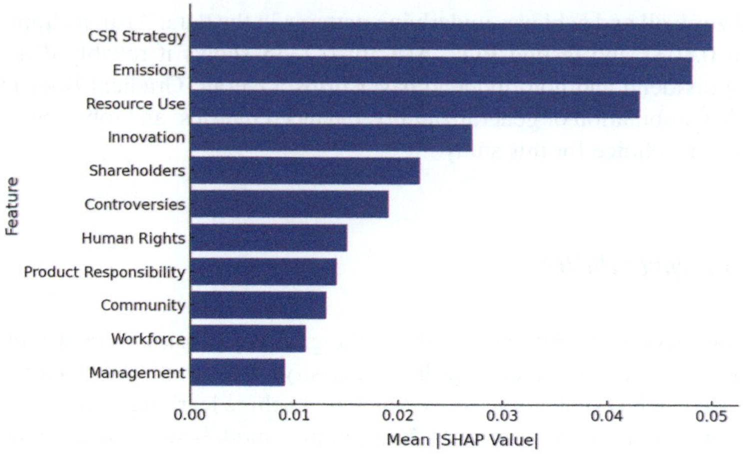

Fig. 1 Feature importance

moderate impact on the model's predictions. The remaining features have a lower impact on the model output.

The SHAP values of the top three features (namely CSR strategy, emissions, and resource use) range between 0.04 and 0.05 which may be observed as small numbers but considering this is a binary classification model, they suggest that they have a significant contribution to the model's prediction. In a binary classification model, SHAP values represent the contribution of each feature to the model's confidence in assigning a firm to one of the two classes. A mean absolute SHAP value of 0.05 for CSR Strategy suggests that, on average, CSR Strategy affects the model's probability prediction by 5 percentage points. This is substantial in a binary context, as small shifts can significantly change the classification outcome, especially when probabilities are close to the decision threshold (typically 0.5).

Figure 2 displays the SHAP summary plot. The mean absolute SHAP values on the X-axis indicate how much each feature can shift the model's prediction. Positive SHAP values push the prediction towards a positive class (e.g., "yes" for dividend payout), while negative values push it towards a negative class (e.g., "no" for dividend payout).

The most important feature of CSR strategy is categorized under the governance pillar, which includes corporate social responsibility strategy, ESG reporting, and firm transparency. It considers to what extent firms combine their social and environmental concerns into their day-to-day decision-making processes. Resource use is related to water, energy, sustainable packaging, and environmental supply chain themes. It points out firms' ability to find eco-friendly solutions by reducing their material use in their production process. The emission category is the indicator of firms' engagement in reducing CO_2 emissions in their production systems. It covers the themes of CO_2 emissions, waste, biodiversity, and environmental management systems (LSEG Data and Analytics 2023).

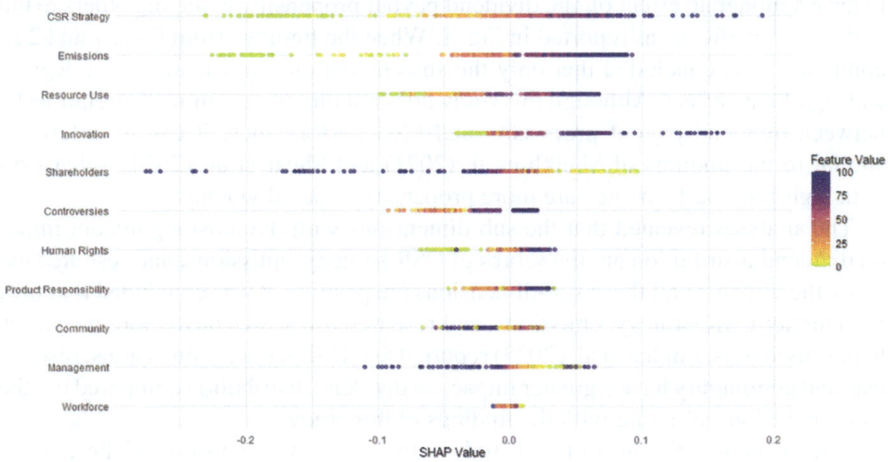

Fig. 2 The SHAP summary plot

The SHAP summary plot in Fig. 2 also provides an interpretation of the gradient boosting model's predictions for dividend payout. Each dot represents a data point in the training data and its SHAP value on the x-axis shows how much that specific feature contributes to the model's prediction for that data point. If a firm has a positive SHAP value for a specific feature, it means that that feature increases the likelihood of paying dividends to that firm and if it has a negative value it means it decreases the likelihood. The color of the dots represents the feature value of the firm with purple indicating higher and yellow representing lower value. So, when we analyze the graph we conclude that as the firms have increased CSR strategy, emissions, and resource use scores the likelihood of paying dividends also increases. For human rights, product responsibility, controversies, community, management, and workforce categories most of the SHAP values are clustered around zero, meaning they do not have any significant impact on the model's predictions. One interesting finding is about the shareholder's category. The firms having the higher shareholders scores have negative SHAP values, indicating that increased shareholders score decreases the likelihood of paying dividends.

5 Conclusion

This study examines the magnitude and direction of the impact of ESG sub-dimensions on a firm's propensity to pay dividends. As reported in Fig. 2, the analyses revealed that the effects of the shareholder, management, community, and workforce sub-dimensions on the propensity to pay dividends are either positive or non-linear. However, among these sub-dimensions, only the shareholder dimension was found to have a moderate effect on the dividend payout propensity, while the others exhibited very low effects, as reported in Fig. 1. When the findings from Figs. 1 and 2 are combined, it is concluded that only the shareholder sub-dimension has a negative and significant effect. Although this study does not directly examine the relationship between dividend payout propensity and ESG performance, it can be inferred—similar to the findings of Sheikh et al. (2021) and Shear et al. (2024)—that firms with high ESG performance are more propensity to pay dividends.

The analyses revealed that the sub-dimensions with the most significant impact on dividend distribution are the scores of CSR strategy, emissions, and resource use. Since the effects of all these sub-dimensions are positive, it was concluded that firms with higher CSR strategy, emissions, and resource use scores have more propensity to pay dividends. Lakhal et al. (2023) reported that CSR strategy, emissions, resource use, and community have a greater impact on dividend distribution compared to other sub-dimensions, aligning with the findings of this study.

One of the key findings of this study is that the sub-dimensions of the environmental pillar—emissions, resource use, and innovation—have a greater impact on the propensity to pay dividends compared to the sub-dimensions of other pillars. Consequently, it was concluded that firms with higher environmental pillar scores are more likely to distribute dividends. The sub-dimensions of the social pillar were found to have the lowest impact on the propensity to pay dividends. The presence of an effect from the environmental pillar, alongside the absence of an effect from the social pillar, is consistent with the findings of Shear et al. (2024). Additionally, Lakhal et al. (2023) reported that the environmental pillar has a greater impact on dividend payout compared to other pillars.

Among the sub-dimensions of governance, while CSR strategy and investor sub-dimensions have a relatively high impact on the propensity to pay dividends, the direction of their effects differs. The impact of other sub-dimensions is relatively minimal. Therefore, discussing a clear relationship between governance pillar performance and the propensity to pay dividends is not feasible. However, an intriguing finding is that the shareholders' sub-dimension, which has a relatively significant impact on the propensity to distribute dividends, negatively influences this propensity. This negative effect, also reported by Lakhal et al. (2023), can be explained through signaling theory. Since investors are, as expected, return-oriented, they are most concerned with the shareholders' sub-dimension among the ESG sub-dimensions. Firms with a low shareholders score may prefer to distribute higher dividends to reduce information asymmetry and agency costs. Conversely, firms that already protect investors' rights may prefer to distribute lower dividends.

In summary, the results show that firms prioritizing ESG are more propensity to pay dividends. This finding indicates that firms with high ESG performance meet the expectations of stakeholders concerned with sustainability through their ESG investments, as well as the expectations of investors who seek returns through dividend distribution. In other words, it can be concluded that the interests of shareholders and other stakeholders do not conflict and that ESG performance (excluding the shareholder's sub-dimension) and dividend distribution decisions function as complementary signals.

This study has several limitations that provide opportunities for future research. First, the analysis focuses on European publicly traded firms, which may limit the generalizability of the findings to other geographical regions with different regulatory environments and ESG practices. Second, the study relies on ESG scores provided by the LSEG database, which, despite its widespread academic use, may not capture all nuances of ESG performance due to variations in reporting standards and data collection methods. Finally, the use of machine learning models, while effective for classification, may obscure certain nuanced relationships due to their complexity, even with SHAP values employed for interpretability.

Future research could examine the interaction effects among ESG sub-dimensions to better understand their combined impact on dividend propensity. The incorporation of ESG metrics could help better understand the relationship between ESG performance and dividend propensity.

References

Ananzeh H, Shbail MOA, Alshurafat H, Amosh HA (2024) A cross-country examination of the relationship between cash holding, dividend policies, and the moderating role of ESG ratings. Cogent Bus Manag 11(1):1–24. https://doi.org/10.1080/23311975.2023.2300523

Benlemlih M (2019) Corporate social responsibility and dividend policy. Res Int Bus Financ 47:114–138. https://doi.org/10.1016/j.ribaf.2018.07.005

Berg F, Fabisik K, Sautner Z (2021) Is history repeating itself? The (un) predictable past of ESG ratings. SSRN Electron J, 1–59

Bhattacharya S (1979) Imperfect information, dividend policy, and "the bird in the hand" fallacy. Bell J Econ 10:259–270. https://doi.org/10.2307/3003330

Bilyay Erdogan S, Ozturk Danisman G, Demir E (2023) ESG performance and dividend payout: a channel analysis. Finance Res Lett 55:1–12. https://doi.org/10.1016/j.frl.2023.103827

Biondi L, Bracci E (2018) Sustainability, popular and integrated reporting in the public sector: a fad and fashion perspective. Sustainability 10(9):1–16. https://doi.org/10.3390/su10093112

Cheung A, Hu M, Schwiebert J (2018) Corporate social responsibility and dividend policy. Account Finance 58(3):787–816. https://doi.org/10.1111/acfi.12238

Coulombe Goulet P, Rapach D, Schütte ECM, Schwenk-Nebbe S (2024) The anatomy of machine learning-based portfolio performance (SSRN Scholarly Paper No. 4628462). Soc Sci Res Netw. https://doi.org/10.2139/ssrn.4628462

Dahiya M, Singh S, Chaudhry N (2023) Corporate social responsibility and dividend policy in India. Manag Decis 61(10):3144–3168. https://doi.org/10.1108/MD-08-2022-1152

Del Vitto A, Marazzina D, Stocco D (2023) ESG ratings explainability through machine learning techniques. Ann Oper Res. https://doi.org/10.1007/s10479-023-05514-z

Dimitropoulos P, Koronios K (2021) Corporate environmental responsibility, cash holding, and dividend policy decisions. In: Dimitropoulos P, Koronios K (eds) Corporate environmental responsibility, accounting, and corporate finance in the EU. Springer, Cham, pp 177–196

Easterbrook FH (1984) Two agency-cost explanations of dividends. Am Econ Rev 74(4):650–659

Ellili NOD (2022) Impact of environmental, social and governance disclosure on dividend policy: what is the role of corporate governance? Evidence from an emerging market. Corp Soc Responsib Environ Manag 29(5):1396–1413. https://doi.org/10.1002/csr.2277

Freeman RE (1984) Strategic management: a stakeholder approach. Pitman, Boston

Hasan MM, Habib A (2020) Social capital and payout policies. J Contemp Account Econ 16(1):1–23. https://doi.org/10.1016/j.jcae.2020.100183

Hastie T, Friedman J, Tibshirani R (2001) The elements of statistical learning. Springer, New York. https://doi.org/10.1007/978-0-387-21606-5

Hussainey K, Walker M (2009) The effects of voluntary disclosure and dividend propensity on prices leading earnings. Account Bus Res 39(1):37–55. https://doi.org/10.1080/00014788.2009.9663348

Jensen MC (1986) Agency cost of free cash flow, corporate finance, and takeovers. Am Econ Rev 76(2):323–329

Kartal MT, Taşkın D, Shahbaz M, Kılıç Depren S, Pata UK (2024) Effects of environment, social, and governance (ESG) disclosures on ESG scores: investigating the role of corporate governance for publicly traded Turkish companies. J Environ Manage 368:1–11. https://doi.org/10.1016/j.jenvman.2024.122205

Kuhn M, Johnson K (2013) Applied predictive modeling. Springer, New York. https://doi.org/10.1007/978-1-4614-6849-3

Lakhal F, Kuzey C, Uyar A, Karaman AS (2023) The relationship between dividend payout and corporate social responsibility: the moderating effect of shareholder friendliness and board monitoring. J Clean Prod 394. https://doi.org/10.1016/j.jclepro.2023.136297

Li X, Xiong H, Li X, Wu X, Zhang X, Liu J, Bian J, Dou D (2022) Interpretable deep learning: Interpretation, interpretability, trustworthiness, and beyond. Knowl Inf Syst 64(12):3197–3234. https://doi.org/10.1007/s10115-022-01756-8

Limkriangkrai M, Koh S, Durand RB (2017) Environmental, social, and governance (esg) profiles, stock returns, and financial policy: Australian evidence. Int Rev Financ 17(3):461–471. https://doi.org/10.1111/irfi.12101

LSEG Data & Analytics (2023) Environmental, social and governance scores from LSEG. https://www.lseg.com/en/data-analytics/sustainable-finance/esg-scores

Lundberg SM, Lee S-I (2017) A unified approach to interpreting model predictions. Adv Neural Inf Process Syst 30. https://proceedings.neurips.cc/paper/2017/hash/8a20a8621978632d76c43dfd28b67767-Abstract.html

Maquieira CP, Espinosa-Mendez C, Arias JT (2024) The impact of environmental, social and governance (ESG) score on dividend payment of large family firms: what is the role of financial constraints? International evidence. Corp Soc Responsib Environ Manag 31(3):2311–2332. https://doi.org/10.1002/csr.2696

Matos PV, Barros V, Sarmento JM (2020) Does ESG affect the stability of dividend policies in Europe? Sustainability 12(21):1–15. https://doi.org/10.3390/su12218804

Matuszewska-Pierzynka A, Mrzyglod U, Pieloch-Babiarz A (2023) ESG performance and dividend stability of the world's largest enterprises. J Entrepreneurship Manag Innovation 19(4):184–217. https://doi.org/10.7341/20231946

Ni X, Zhang H (2019) Mandatory corporate social responsibility disclosure and dividend payouts: evidence from a quasi-natural experiment. Account Finance 58(5):1581–1612. https://doi.org/10.1111/acfi.12438

Raschka S (2020) Model evaluation, model selection, and algorithm selection in machine learning (No. arXiv:1811.12808). arXiv. http://arxiv.org/abs/1811.12808

Russell SJ, Norvig P (2016) Artificial intelligence: a modern approach. Pearson. https://thuvienso.hoasen.edu.vn/handle/123456789/8967

Saeed A, Zamir F (2021) How does CSR disclosure affect dividend payments in emerging markets? Emerg Mark Rev 46:1–20. https://doi.org/10.1016/j.ememar.2020.100747

Salah OB, Amar AB (2022) Does corporate social responsibility affect dividend policy? Empirical evidence in the French context. J Glob Responsib 13(3):268–289. https://doi.org/10.1108/JGR-10-2021-0082

Salvi A, Nirino N, Battisti E, Gianfrancesco I (2024) Payout policy and ESG: a European investigation. Res Int Bus Financ 68:1–10. https://doi.org/10.1016/j.ribaf.2023.102189

Shapley LS (1953) A value for n-person games. In: Contribution to the theory of games, 2. https://www.torrossa.com/gs/resourceProxy?an=5641636&publisher=FZO137#page=87

Shear F, Rizwan MS, Ashraf D (2024) ESG orientation of firms and dividend payout: an international investigation. Retrieved from https://ssrn.com/abstract=4992183. https://doi.org/10.2139/ssrn.4992183

Sheikh MF, Bhutta AI, Rehman B, Bazil M, Hassan A (2021) Corporate social responsibility hand dividend policy: a strategic choice in family firms. J Family Bus Manag 12(2):296–315. https://doi.org/10.1108/JFBM-10-2020-0096

Trihermanto F, Yunieta AN (2020) Corporate life cycle, CSR, and dividend policy: empirical evidence of Indonesian listed firms. Social Responsib J 16(2):159–178. https://doi.org/10.1108/SRJ-09-2017-0186

Zadeh MH (2021) The effect of corporate social responsibility transparency on corporate payout policies. Int J Manag Finance 17(5):708–732. https://doi.org/10.1108/IJMF-07-2020-0386

Zahid RMA, Taran A, Khan MK, Chersan IC (2023) ESG, dividend payout policy and the moderating role of audit quality: empirical evidence from Western Europe. Borsa Istanbul Rev 23(2):350–367. https://doi.org/10.1016/j.bir.2022.10.012

Machine Learning Algorithms to Study the Impact of Sustainability on Financial Success: Evidence from US Stock Market

Merve Doğruel and Haydar Anıl Küçükgöde

Abstract In recent years, as in almost every field, the concept of sustainability has become a concept that has been given considerable importance and researched in both the academic and business worlds in the field of finance. This study aims to find the best model for estimating financial success within the framework of sustainability by adding environmental risk score, social risk score, and governance risk score in addition to classical indicators in stock returns. For this purpose, data obtained from 300 US companies across 11 distinct industries is utilized. The features evaluated are the environmental risk score, social risk score, governance risk score, as well as profitability, liquidity, leverage, RoA, and beta, which are used in classical studies. To obtain the best modeling predictions, various machine learning algorithms were used instead of classical statistical methods. Based on the F1 performance metrics of the seven machine learning algorithms tested, the model with the highest performance is Random Forest, an ensemble learning model. Based on the Random Forest model, environmental and social risk scores are particularly important features for financial success.

Keywords Machine learning · Random forest · Sustainable finance · Environmental score · Social score · Governance score

M. Doğruel (✉)
Department of Management Information Systems, Faculty of Business and Management Sciences, Istanbul Esenyurt University, İstanbul, Türkiye
e-mail: mervedogruel@esenyurt.edu.tr

H. A. Küçükgöde
Department of International Trade and Finance, Faculty of Business and Management Sciences, Istanbul Esenyurt University, İstanbul, Türkiye
e-mail: haydaranilkucukgode@esenyurt.edu.tr

M. Gün and B. Kartal (eds.), *Machine Learning in Finance*, Contributions to Finance and Accounting, https://doi.org/10.1007/978-3-031-83266-6_8

1 Introduction

In recent decades, environmental risks arising from climate change compelled policymakers to prioritize sustainability practices. In line with that, Corporate Social Responsibility (CSR) has become a significant concern for companies, not only in terms of environmental concerns but also to attract investors who are more concerned about the expenditure of their funds. A new reporting system was introduced due to increasing awareness among investors, customers, suppliers, and all stakeholders about risks associated with the environment, social responsibility, and governance. Companies report their risks under three categories: Environmental, Social, and Governance (ESG).

ESG initiatives have assumed a crucial role in the expanding realm of corporate sustainability, and the proactive adoption of ESG principles by firms is increasingly prevalent. Consequently, the volume of investments geared towards ESG has experienced a significant surge in recent years. Currently, ESG serves as the primary criterion for assessing and guaranteeing a company's compliance with sustainability standards (Howard-Grenville 2021). This study questions whether ESG risk scores have an impact on stock returns of US companies by examining how environmental, social, and governance risk scores are separately and individually associated with returns.

The stakeholder theory asserts that employers, customers, suppliers, and the community are among the stakeholders that companies are accountable for their actions (Freeman 2010). Business an association that comprises more than 200 chief executive officers (CEOs) from the most prominent companies in the US released a statement in August 2019 and announced that the purpose of a company should be to consider the benefits of not only shareholders but all stakeholders including the environment and communities (Business Roundtable Statement 2019).

ESG principles are increasingly being integrated into business practices. As a result, ESG-focused investments have experienced exponential growth. At the onset of 2020, the Global Sustainable Investment Alliance reported that the aggregate value of sustainable investments in advanced economies had surpassed $35 trillion (Global Sustainable Investment Alliance 2020). The significance of ESG factors in investment decisions has increased that much lately, owing to the escalating interest of both investors and regulators in socially responsible investments and impact finance. Even though sustainability has been a subject of academic interest since the 1970s, the rapidly expanding field of ESG has experienced a significant surge in academic attention. However, the literature regarding the correlation between ESG scores and company value or stock returns is inconclusive. Some studies indicate that there is a positive relation between ESG scores and company value (Buchanan et al. 2018; Li et al. 2018; Xie et al. 2019; Giese et al. 2019; Qureshi et al. 2020; Yi and Demirel 2023; Seok et al. 2024) while others suggest that there is no discernible relationship or no relationship at all (Duque-Grisales and Aguilera-Caracuel 2021; Horváthová 2010).

Stakeholders of companies are increasingly relying on ESG metrics to assess the ethical commitment and long-term commitment to the sustainability of those companies (Alkaraan et al. 2024) which supports the significant benefits of ESG information disclosures' impact on financial performance (Chen and Xie 2022). Tan and Zhu (2022) present a novel viewpoint on conventional financial risk analysis, emphasizing that ESG ratings not only foster green innovation but also alleviate financial constraints. They indicate that companies with high ESG scores are more likely to receive favorable financing terms, especially during growth stages, and are more committed to sustainable projects.

In this study, authors obtain ESG scores from Sustainalytics and assess how companies handle environmental, social, and governance risks that can affect their financial health. These scores help investors gauge the impact of these factors on a company's performance and long-term sustainability. By categorizing risk levels from minimal to severe across diverse industries worldwide, these scores serve as a valuable tool for investors seeking responsible investment, product development, and sustainable portfolio management.

In addition to ESG factors, conventional indicators, which have been widely studied in the literature to explain stock returns, are also included in the models employed in this study.

Markowitz's (1952) groundbreaking Modern Portfolio Theory shows how risk-averse investors can create optimal portfolios that maximize expected returns for a given level of risk. Sharpe (1964) and Lintner (1965) carried Markowitz's theory one step forward by developing the Capital Asset Pricing Model (CAPM). CAPM is a model where beta, the coefficient of the market risk premium, shows how much systematic risk there is. That beta indicates the sensitivity of a stock's excess return to that of the market. Hence, beta has been an essential indicator of stock/portfolio returns for nearly sixty years, and it is included in our model. The CAPM can be expressed as follows:

$$\mathrm{r_e} = \mathrm{r_f} + \beta.(\mathrm{E}(\mathrm{r_m}) - \mathrm{r_f}) \tag{1}$$

where r_f stands for the risk-free interest rate, r_e is the required rate of return by shareholders, β indicates the systematic risk, and $\mathrm{E}(\mathrm{r_m})$ represents the expected return on the market portfolio.

Despite maintaining its place in both practice and theory, CAPM has been one of the most scrutinized and tested theories in the history of finance. Fama and French's (1992) three-factor model includes two factors over CAPM's market risk premium and takes the impact of financial leverage on stock return into account. There exist numerous studies that establish a significant relationship between stock returns and financial leverage (Matemilola et al. 2013; Cornaggia et al. 2019; Doshi et al. 2019; Gomes and Schmid 2021; Do et al. 2022). The leverage ratio is an additional indicator utilized in this study to estimate stock returns.

Another financial ratio that will take place in the models is liquidity. Stocks of less liquid companies perform better than liquid ones in the US stock market, as well as other developed and emerging markets (Pástor and Stambaugh 2003; Liu

2006; Bekaert et al. 2007; Lee 2011; Amihud 2014). Chiang and Zheng (2015) find that excess stock returns of the G7 countries exhibit a negative relationship with illiquidity at the firm level. Jihadi et al. (2021) show that the liquidity ratio has a significant effect on firm value. Marozva (2019) adds liquidity as the fourth factor to the Fama–French three-factor model and finds that it is a significant indicator of excess stock returns.

In addition to the three pillars of ESG, stock beta, leverage, and liquidity, profitability is the next variable to explain stock performance. Scholars have demonstrated that profitability possesses predictive power for excess stock returns (Hou and Zhang 2015; Chue and Xu 2022). Alaagam (2019) finds no long-term relationship between profitability and stock prices, but a significant association in the short term. Jihadi et al. (2021) demonstrate that profitability ratio has a significant effect on firm value. Berggrun et al. (2020) show that the positive impact of profitability on stock returns extends to subregional markets, small and large stocks.

Return on assets (RoA) assesses a firm's efficiency in utilizing its assets to generate revenue and therefore is crucial for evaluating asset utilization across all businesses (Diaz and Pandey 2019). Jihadi et al. (2021) find a positive effect of RoA on firm value. Scholars support the theory that RoA has a significant positive effect on stock returns (Loughran and Wellman 2011; Nadyayani and Suarjaya 2021). RoA represents the last variable of the models to be employed in this study to explain the stock returns.

This research aims to examine the incremental explanatory power of ESG variables on stock returns, controlling for the well-established risk factors beta and leverage, as employed in the CAPM and Fama–French frameworks.

2 Data

To investigate the significance of the individual effects of each pillar of the ESG on stock returns, and environmental, social, and governance risk scores of 300 US companies 11 different sectors have been obtained from Sustainalytics.[1]

Daily prices of the stocks of 300 US companies and S&P 500 index levels between October 2023 and October 2024 are provided by Refinitiv. For each stock, the authors calculated the differences between the average stock return and the market (index) return. The results have been favorable for 112 stocks, indicating that they have delivered higher returns than the market index, whereas the remaining 188 stocks have maintained a lower return than the market index. In Sect. 5, these positive (successful) and negative (unsuccessful) differences indicate the measure of financial success.

Stock betas are obtained from Yahoo! Finance, which are calculated based on 5-year monthly data. The beta of each stock is determined by dividing the covariance between stock returns and the market index by the variance of returns on the market

[1] https://www.sustainalytics.com/esg-ratings.

index. The average beta of the sample is 1.04, where 154 companies have betas less than 1, four have equaled 1, and 142 companies have greater betas than 1.

Financial ratios, namely leverage, liquidity, profitability, and return on assets are calculated by the authors on company financials.

The leverage ratio is a measure of a company's debt-to-equity ratio, which measures the indebtedness of a company relative to its shareholder equity. It is calculated by taking the sum of both short-term and long-term debt and dividing it by total equity.

The current ratio (CR) is the most commonly utilized liquidity ratio in both academic literature and practice. It is simply a measure of the ability of a company to overcome its short-term obligations and is assessed by dividing current assets by short-term liabilities. Values below 1 indicate some liquidity issues for companies. The liquidity ratio has been averaged at 1.62, where 87 companies have numbers below 1, 3 have equal to 1, and 210 companies have greater than 1. Half of those 210 companies have CRs above 1.5, which means the value of their liquid assets is at least 50% higher than their short-term obligations.

Operating profit margin is generally considered a more reliable profitability ratio than net profit margin. It solely focuses on a company's fundamental operations, excluding factors such as interest expenses, taxes, and non-operating gains or losses. This provides a clearer picture of how efficiently a company manages its day-to-day activities. Since the operating profit margin excludes non-operating factors, it is less vulnerable to fluctuations caused by external factors, such as changes in tax rates or interest rates. This renders it a more consistent and dependable indicator of profitability. The calculation is performed by dividing the operating profit by the total revenue.

The return on assets (RoA) ratio gives valuable insight into a company's asset utilization by comparing net income to average total assets. It is considered a measure of management efficiency because it depends on how well the resources of the company are allocated.

Table 1 depicts the condensed dataset. The dataset comprises 300 observations (companies), eight attributes (profitability, liquidity, leverage, RoA, beta, environmental risk score, social risk score, governance risk score), and one outcome (success).

3 Methodology

To make this prediction, the most popular supervised classification algorithms, such as Logistic Regression, Classification and Regression Tree (CART), K-Nearest Neighbors (KNN), Random Forest (RF), Bagging, and Support Vector Machine (SVM), will be used.

Table 1 Condensed dataset

Company	Profitability	Liquidity	Leverage	RoA	Beta	ERS	SRS	GRS	Success
A. O. Smith	20.24	1.70	0.09	14.53	1.15	7.2	11.9	6.4	0
Amazon	9.92	1.10	0.67	6.58	1.15	6.3	14.8	7.9	1
Berkshire H	41.77	2.28	0.20	5.08	0.87	1	4.9	14.8	0
Nvidia	62.06	4.27	0.17	55.26	1.67	2.4	4.7	6	1
Warner Bros	− 6.71	0.76	1.16	− 0.12	1.49	2.3	9.2	6.7	0
Waste M	19.75	1.07	2.25	7.88	0.75	8.1	5.2	3.3	1

3.1 Logistic Regression

Logistic regression is a classification algorithm that utilizes linear regression to evaluate output and reduce error. Logistic regression models are classification models used specifically to distinguish between two different categories (Geetha and Sendhilkumar 2023).

The logistical approach to discrimination assumes that the log of the group-conditional densities can be expressed as a linear combination of the input variables as follows (Papageorgiou et al. 2008; Prathom and Sujitapan 2024):

$$\log \frac{p}{1-p} = f(x) = w_0 + w_1x_1 + w_2x_2 + \cdots + w_nx_n \tag{2}$$

The probability of a dichotomy is assumed to follow the cumulative logistic distribution function after incorporating exponentials on both sides:

$$p = \frac{1}{1 + e^{-f(x)}} \tag{3}$$

The w_0 parameter in the model represents the constant, while the w parameter represents the coefficient vector. The estimate is made according to the maximum likelihood procedure.

Logistic regression has a confidence limit of 0 and 1, whereas linear regression models can take any real number (Geetha and Sendhilkumar 2023).

3.2 K-Nearest Neighbors (KNN)

The KNN is a supervised learning classifier that classifies or predicts the grouping of an individual data point based on the distance between distinct feature vectors. KNN, a nonparametric algorithm, performs computations by exploiting the similarity between observations to classify them. The measurement of similarity is frequently referred to as distance, proximity, or closeness (Zheng et al. 2023; Mirtaheri and Shahbazian 2022).

Although the most frequently used distance measure in the algorithm is Euclidean, Manhattan, and Minkowski distances are also frequently used distance measures. The Euclidean distance is the distance between two points in the plane or hyperplane. It is calculated as follows (Gupta et al. 2024):

$$d(x, y) = \sqrt{\sum_{i=1}^{n} (x_i - y_i)^2} \tag{4}$$

wherein x and y are points that represent two points in n-dimensional space.

The parameter k of the KNN has a significant impact on the classification result, and it is attempted to select the optimal k based on the available data. Generally, a larger k reduces the effect of noise on the classification, but the class boundaries are less distinct (Zheng et al. 2023; Mirtaheri and Shahbazian 2022). It is advisable to select an odd value for k to prevent ties in classification. Cross-validation methods can help select the best k value for a given dataset (Gupta et al. 2024).

The steps of the algorithm may be categorized as follows (Mirtaheri and Shahbazian 2022):

- Normalize the features to have a mean of zero and a variance of one.
- Set K to your chosen number of neighbors.
- For each new observation, calculate the distance between the training examples and the current observation, and then add the distance and index of the example to an ordered collection.
- Sort the distances and indices in ascending or descending order.
- Choose the initial K entries from the sorted collection.
- Obtain the labels of the K entries that have been selected.
- If classification is concerned, return the mode of the K labels.

3.3 *Classification and Regression Tree (CART)*

Decision tree algorithms produce predictions based on repeated binary separations, which are non-parametric and non-linear. It is utilized to generate both classification and regression trees (Doğruel and Soner Kara 2023; Doğruel and Fırat 2021).

A set of if–then rules is created in all methods used to construct decision trees, resulting in a final set of values for the variable to be predicted. If the final values are the probabilities of a categorical variable, the decision tree created is called a classification tree. If the quantities of a continuous variable are the final values obtained, the decision tree created is called a regression tree (Doğruel and Fırat 2021).

The fundamental steps of the CART algorithm are as follows (Koçyiğit 2023):

- Start with the node that decides.
- Finds the split among the available candidates that reduces impurity to its purest extent.
- Utilize the split technique to divide the decision node into two nodes, commonly referred to as interior node.
- Repeat the three-step search until the tree meets its requirements.

The CART algorithm employs a binary decision tree. During each step of the tree-building process, the rule generated within the node divides the specified sample set into two distinct parts: the portion where the rule is executed (the right subtree) and the portion where it is not executed. This method examines all possible ways to branch for each node and chooses the one with the greatest improvement. The estimation algorithm employed in the CART algorithm is founded on the intuitive concept of reducing the uncertainty (heterogeneity) in the node and the Gini impurity index (Karminsky and Morgunov 2021):

$$\mathrm{Gini(T)} = 1 - \sum_{i=1}^{n} p_i^2 \tag{5}$$

where p_i is the probability of class i being in T.

If T is divided into two parts, T1 and T2, with the number of examples in each N1 and N2, respectively, then the split quality indicator is equal to:

$$\mathrm{Gini}_{\mathrm{split}}^{(T)} = \frac{N1}{N}\mathrm{xGini(T1)} + \frac{N2}{N}\mathrm{xGini(T2)} \tag{6}$$

The lowest $\mathrm{Gini}_{\mathrm{split}}^{(T)}$ is the best partition.

3.4 Random Forest (RF) and Bagging

Tian Jinhao created the first random forest algorithm in 1995, using the random subspace method. Breiman and Cutler devised an extension of this algorithm, and the term “Random Forest” became their trademark (Li 2022).

Random forest, a widely used form of ensemble learning, is a machine learning technique in which each tree is constructed using randomly selected subsets of both data and features. The random forest technique works by training individual trees on random vectors sampled independently from the same distribution. Random forest is an integrated learning approach for regression, classification, and other tasks that utilizes a substantial number of decision trees during training, resulting in the generation of categories as category patterns or average predictions of individual trees (Li 2022). The decrease in generalization error, which pertains to anticipating data that has not been previously observed, is correlated with the increase in the number of trees present in the forest (Breiman 2001).

The Random Forest operates on the principle of bagging. The Bagging algorithm utilizes a collective learning approach to construct a classifier group. In bagging, the given training data is divided into smaller training datasets by employing the sampling method. The Bagging algorithm aims to provide diversity by training each basic learning algorithm on different training sets. The straightforward random substitution sampling technique is commonly employed to generate diverse training sets from the data set. A majority vote is used to merge the results of the training sets from the sampling method and the classification methods trained, using the majority vote (Doğruel and Soner Kara 2023; Sumathi et al. 2022; Onan 2018).

In a random forest model, the error rate is dependent on two factors (Jaiswal and Samikannu 2017):

- The error rate will only increase as the correlation between two trees within the forest increases.
- A tree's strength is determined by its low error rate. The lower the error rate, the stronger the tree will be.

The feature variables selected for splitting at an internal node are formulated as an optimization problem based on certain splitting criteria, whether it is a classification or regression problem. Entropy is a common splitting criterion in classification problems, which is a practical application of the source coding theorem. This theorem determines the upper limit of the bit representation of a random variable. At each internal node, the entropy is determined by the formula (Schonlau and Zou 2020):

$$E = -\sum_{i=1}^{c} p_i \times \log(p_i) \tag{7}$$

where c is the number of distinct classes, and pi is the prior probability of each particular class. This value is maximized to gain the maximum information at every split of the decision tree. For regression problems, a widely employed splitting criterion is the mean squared error at each internal node.

Numerous studies on random forest employ the default size of the candidate feature set as $m \approx \sqrt{p}$ in classification and $m \approx p/3$ in regression problems (Doğruel and Soner Kara 2023).

The basic properties of random forest can be stated as follows (Jaiswal and Samikannu 2017):

- It has been regarded as an unmatched algorithm in terms of accuracy.
- Even in large data sets with hundreds or thousands of input variables, there is no overfitting problem and no data pruning is required.
- It is utilized for the selection of feature subsets and the imputation of missing data, and it performs admirably.
- During the forest construction process, the random forest algorithm generates an internal unbiased estimate of the generalization error.
- The constructed forest possesses the capability to effectively incorporate data in the future.

3.5 Support Vector Machine (SVM)

Support Vector Machine (SVM) creates the best segmentation line that can divide the n-dimensional space into classes. High-dimensional datasets are then classified in an optimal decision plane. This optimal plane is called a hyperplane. As SVM relies on the logic of selecting extreme foci/vectors that aid in the formation of the hyperplane, these extreme cases are referred to as support vectors, and the computation is referred to as a Support Vector Machine (Kurani et al. 2023; Zhang et al. 2021).

Consider the following training set (Kalita et al. 2020)

$$\left\{\left(X_i, y_i\right),\ i = 1 \ldots n\right\},\ X_i \in \mathbb{R}^m, y_i \in \{+1, -1\} \tag{8}$$

If the training set is linearly separable, there exists w $\in \mathbb{R}^m$ and b $\in \mathbb{R}$ such that.

$$\begin{aligned} w^T X_i + b > 0 \forall_i, \quad \text{s.t. } y_i = +1 \\ w^T X_i + b < 0 \forall_i, \quad \text{s.t. } y_i = -1 \end{aligned} \tag{9}$$

Here $w^T X_i + b = 0\forall_i$ will be the separating hyperplane. The hyperplane chosen should maximize the margin between the two classes, even if there are many hyperplanes separating instances of two classes.

The kernel trick is a widely used method for dealing with nonlinear separable problems in SVM. The standard multivariate version of the kernel trick requires separating the data from both populations using a symmetric, non-negative, definite kernel function. The kernel values can be regarded as the inner product of transformed versions of the initial observations in a different (usually higher-dimensional) space. This circumstance can be deemed equivalent to the creation of a nonlinear classifier within a low-dimensional space. It is expected that the groups will be able to distinguish themselves better in the new finals (Cárcamo et al. 2024; Kalita et al. 2020).

4 Models and Empirical Findings

This study aims to estimate the financial success output, which has two levels: successful and unsuccessful (as explained in Sect. 2). For this purpose, financial sustainability risk indices are added to the study as a feature, in addition to traditional features. Furthermore, to make a classification using the best model, various machine learning algorithms were used instead of classical statistical methods.

Seven algorithms were tried to estimate this categorical variable with two levels, including logistic regression, CART, KNN, Bagging, Random Forest, SVM Linear, and SVM Nonlinear. These algorithms are the most popular supervised classification algorithms in the literature. All of these algorithms are executed utilizing R Programming.

Data from a total of 240 companies, comprising 80% of the 300 companies, is utilized for the training of the tested machine learning algorithms. The remaining 20% of company data (60 companies) is used for testing.

The F1 metric of the training data in the Logistic Regression algorithm applied with tenfold cross-validation during the tuning process is 0.7674, while the F1 metric of the test data is 0.6585.

While creating the model in the CART algorithm, the min split argument, which represents the number of observations at each node, is tested between 5 and 25 in the tuning process, and the cp argument, which determines the complexity of the model is tested at 0.01 intervals from 0.03 to 0.05. After the tuning procedure, the optimal model is obtained at values where the minimum split argument is 5 and the maximum split argument is 0.05. In the best model, where the risk feature is the root node, the F1 metric for training data is calculated to be 0.8218, while the F1 metric for test data is calculated to be 0.6197.

The KNN algorithm is trained with tenfold cross-validation, and k values are tried between 1 and 20 in the tuning process. The best k value is determined at the end of the tuning process to be 2. The F1 metric for the training data of the best model is 0.7988, whereas the F1 metric for the test data is 0.6988.

In the model obtained with the bagging algorithm, the F1 metric of the training data is calculated to be 1, and the F1 metric of the test data is calculated to be 0.7013. The tuning procedure is utilized to evaluate the random forest algorithm. The mtry argument, which specifies the number of independent features to be selected for the creation of trees in the model, is subjected to testing ranging from 2 to 8, and a tenfold cross-validation is employed. It has been observed that setting the mtry argument to 3 would result in a superior model. The F1 metric of the training data for this model is found to be 1, and the F1 metric of the test data is found to be 0.7179. The F1 metric of the training data is found to be 1, without the tuning process, while the F1 metric of the test data is found to be 0.7407. Figure 1 shows the ranking of the features of the created RF model according to the mean decrease in accuracy and the mean decrease in gini measurements. This information aids us in selecting variables. It is, in other words, possible to identify the most important variables in the model.

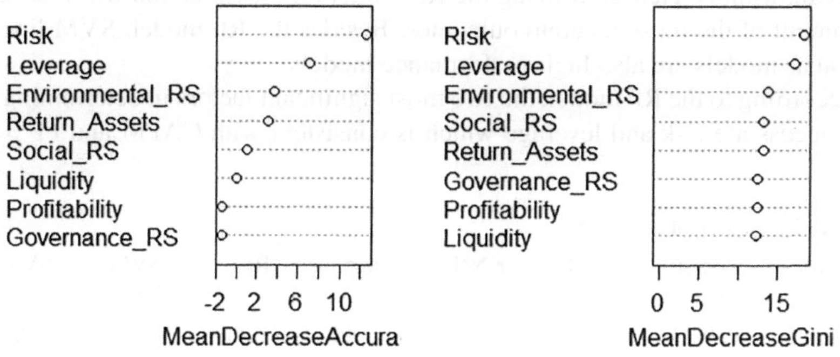

Fig. 1 Variable importance plot

When looking at Fig. 1, it is clear that the environmental sustainability indicator and then the social sustainability indicator are important indicators for predicting the financial success of businesses.

The SVM algorithm has been implemented to construct both a linear and a nonlinear model. The cost argument, which is utilized to establish a balance between error tolerance and model complexity in the linear model, is tuned. At the end of the tuning process, the cost argument takes the value of 10 as the value of 10. The F1 score of this model for the training data is 0.7805, whereas the F1 score for the test data is 0.7391. To generate the nonlinear model, the gamma argument, which determines the domain of the kernel function, is additionally subjected to the tuning process along with the cost argument. As a consequence of performing the tune operation, the cost argument has been set to 1 and the gamma argument has been set to 0.5. The F1 metric for the model's training data is calculated as 0.9034, while the F1 metric for the test data is calculated as 0.7317.

5 Conclusion

This study aims to estimate company success by including the well-known risk factors beta and leverage used in the CAPM and Fama–French frameworks, as well as financial sustainability features. The success cases are labeled according to whether the average stock returns are above or below the average market return. When estimating the best model, it is also important to determine the importance of sustainability features. In this manner, the authors have run seven machine learning models, including three financial ratios (liquidity, profitability, and return on assets), in addition to the environmental risk score, social risk score, governance risk score, beta, and leverage of 300 US companies.

Since there are unbalanced class distributions, the model performances should be interpreted with F1 metrics. The F1 metrics for the seven machine learning models attempted are presented in Table 2. It can be said that the RF algorithm, which gives the highest F1 value (0.7407) for the test data, has the best performance among the tested algorithms. Hence, utilizing the RF model for classification will facilitate the attainment of the most accurate outcomes. Besides the RF model, SVM linear and nonlinear models are also high-performance models.

According to the RF model, the two most significant factors in determining financial success are risk and leverage which is consistent with CAPM and FF 3-factor

Table 2 F1 metrics of algorithms

F1 metrics	Logistik Reg.	CART	KNN	Bagging	RF	SVM linear	SVM nonlinear
Train	0.7674	0.8218	0.7988	1.0000	1.0000	0.7805	0.9034
Test	0.6585	0.6197	0.6988	0.7013	0.7407	0.7391	0.7317

models. When examining the effects of sustainability indicators on financial success, one of the main research questions of the study is that the environmental sustainability indicator ranks right after the risk and leverage attributes (Fig. 1). The social sustainability indicator can help predict financial success, but the governance sustainability indicator is less helpful.

When both MD accuracy and MD gini outputs are considered in Fig. 1, there is a consistency in terms of the five most influential factors and the three factors with relatively low impact.

Future studies have the potential to further validate the algorithms by testing them on a larger dataset that encompasses a broader range of companies from diverse nations, incorporating ESG ratings from other providers, and utilizing time-series analysis. The availability of a larger dataset would also facilitate sector-level analysis, thereby enabling comparative analysis of the impact of ESG pillars on stock returns across various sectors. Furthermore, it is feasible to conduct prediction studies using deep learning models.

References

Alaagam A (2019) The relationship between profitability and stock prices: evidence from the Saudi Banking Sector. Res J Finance Account 10(14). Available at SSRN: https://ssrn.com/abstract=4248045

Alkaraan F, Elmarzouky M, Hussainey K, Venkatesh VG, Shi Y, Gulkz N (2024) Reinforcing green business strategies with Industry 4.0 and governance towards sustainability: natural-resource-based view and dynamic capability. Bus Strategy Environ 33(4):3588–3606. https://doi.org/10.1002/bse.3665

Amihud Y (2014) The pricing of the illiquidity factor's systematic risk. Available at SSRN 2411856. https://doi.org/10.2139/ssrn.2411856

Bekaert G, Harvey CR, Lundblad C (2007) Liquidity and expected returns: lessons from emerging markets. Rev Financ Stud 20(6):1783–1831. https://doi.org/10.1093/rfs/hhm030

Berggrun L, Cardona E, Lizarzaburu E (2020) Firm profitability and expected stock returns: evidence from Latin America. Res Int Bus Financ 51:101119. https://doi.org/10.1016/j.ribaf.2019.101119

Breiman L (2001) Random forest. Mach Learn 45:5–32. https://doi.org/10.1023/A:1010933404324

Buchanan B, Cao CX, Chen C (2018) Corporate social responsibility, firm value, and influential institutional ownership. J Corp Finan 52:73–95. https://doi.org/10.1016/j.jcorpfin.2018.07.004

Business Roundtable Statement (2019). https://www.businessroundtable.org/business-roundtable-redefines-the-purpose-of-a-corporation-to-promote-an-economy-that-serves-all-americans

Cárcamo J, Cuevas A, Rodríguez L (2024) A uniform kernel trick for high and infinite-dimensional two-sample problems. J Multivar Anal 202. https://doi.org/10.1016/j.jmva.2024.105317

Chen Z, Xie G (2022) ESG disclosure and financial performance: moderating role of ESG investors. Int Rev Financ Anal 83:102291. https://doi.org/10.1016/j.irfa.2022.102291

Chiang TC, Zheng D (2015) Liquidity and stock returns: evidence from international markets. Glob Financ J 27:73–97. https://doi.org/10.1016/j.gfj.2015.04.005

Chue TK, Xu JK (2022) Profitability, asset investment, and aggregate stock returns. J Bank Finance 143:106597. https://doi.org/10.1016/j.jbankfin.2022.106597

Cornaggia K, Simin T, Sonmez-Leopold F, Betting on Leverage (2019) Available at SSRN: https://ssrn.com/abstract=3352156

Diaz JF, Pandey R (2019) Factors affecting return on assets of US technology and financial corporations. Jurnal Manajemen Dan Kewirausahaan 21(2):134–144. https://doi.org/10.9744/jmk.21.2.134-144

Do HX, Nguyen NH, Nguyen QM (2022) Financial leverage and stock return co-movement. J Financ Markets 60:100699. https://doi.org/10.1016/j.finmar.2021.100699

Doğruel M, Fırat SÜ (2021) Veri madenciliği karar ağaçları kullanarak ülkelerin inovasyon değerlerinin tahmini ve doğrusal regresyon modeli ile karşılaştırmalı bir uygulama. Istanbul Bus Res 50(2):465–493. https://doi.org/10.26650/ibr.2021.50.015019

Doğruel M, Soner Kara S (2023) Determining the happiness class of countries with tree-based algorithms in machine learning. Acta Infologica 7:243–252. https://doi.org/10.26650/acin.1251650

Doshi H, Jacobs K, Kumar P, Rabinovitch R (2019) Leverage and the cross-section of equity returns. J Financ 74(3):1431–1471. https://doi.org/10.1111/jofi.12758

Duque-Grisales E, Aguilera-Caracuel J (2021) Environmental, social and governance (ESG) scores and financial performance of multilatinas: moderating effects of geographic international diversification and financial slack. J Bus Ethics 168(2):315–334. https://doi.org/10.1007/s10551-019-04177-w

Fama EF, French KR (1992) The cross-section of expected stock returns. J Financ 47(2):427–465. https://doi.org/10.1111/j.1540-6261.1992.tb04398.x

Freeman RE (2010) Strategic management: a stakeholder approach. Cambridge University Press

Geetha TV, Sendhilkumar S (2023) Machine learning: concept, techniques and applications. CRC Press, Boca Raton

Giese G, Lee L-E, Melas D, Nagy Z, Nishikawa L (2019) Foundations of ESG investing: how ESG affects equity valuation, risk, and performance. J Portfolio Manag 45:69–83

Global Sustainable Investment Alliance (2020) Global sustainable investment review. https://www.gsi-alliance.org/wp-content/uploads/2021/08/GSIR-20201.pdf

Gomes JF, Schmid L (2021) Equilibrium asset pricing with leverage and default. J Financ 76(2):977–1018. https://doi.org/10.1111/jofi.12987

Gupta V, Bansal T, Kumar Yadav V, Bhowmik D (2024) Artificial intelligent in indoor or outdoor surveillance systems: a systematic view, principles, challenges and applications. In: Tripathi P, Rai M, Kumar N, Kumar S (eds) Mathematical models using artificial intelligence for surveillance systems. Scrivener Publishing, USA, pp 293–334

Horváthová E (2010) Does environmental performance affect financial performance? A meta-analysis. Ecol Econ 70(1):52–59. https://doi.org/10.1016/j.ecolecon.2010.04.004

Hou K, Xue C, Zhang L (2015) Digesting anomalies: an investment approach. Rev Financ Stud 28(3):650–705. https://doi.org/10.1093/rfs/hhu068

Howard-Grenville J (2021) ESG impact is hard to measure-but it's not impossible. Harvard Bus Rev. https://hbr.org/2021/01/esg-impact-is-hard-to-measure-but-its-not-impossible

Jaiswal JK, Samikannu R (2017) Application of random forest algorithm on feature subset selection and classification and regression. In: 2017 World congress on computing and communication technologies (WCCCT), Tiruchirappalli, India, 2017, pp 65–68. https://doi.org/10.1109/WCCCT.2016.25

Jihadi M, Vilantika E, Hashemi SM, Arifin Z, Bachtiar Y, Sholichah F (2021) The effect of liquidity, leverage, and profitability on firm value: empirical evidence from Indonesia. J Asian Finance, Econ Bus 8(3):423–431. https://doi.org/10.13106/jafeb.2021.vol8.no3.0423

Kalita DJ, Singh VP, Kumar V (2020) A survey on SVM hyper-parameters optimization techniques. In Kumar Shukla R, Agrawal J, Sharma S, Chaudhari NS, Shukla KK (eds) Social networking and computational intelligence: proceedings of SCI-2018. Springer Nature Singopore, Singapore, pp 243–256

Karminsky A, Morgunov A (2021) Bank credit risk modeling in emerging capital markets. In: Karminsky AM, Mistrulli PE, Stolbov MI, Shi Y (eds) Risk assessment and financial regulation in emerging markets' banking. Springer Nature Switzerland AG, Switzerland

Kurani A, Doshi P, Vakharia A, Shah M (2023) A comprehensive comparative study of artificial neural network (ANN) and support vector machines (SVM) on stock forecasting. Ann Data Sci 10:183–208. https://doi.org/10.1007/s40745-021-00344-x

Koçyiğit H (2023) Process of machine learning methods. In: Ahangar RG, Napier M (eds) Advancement in business analytics tools for higher financial performance. IGI Global, USA, pp 1–38

Lee KH (2011) The world price of liquidity risk. J Financ Econ 99(1):136–161. https://doi.org/10.1016/j.jfineco.2010.08.003

Li A (2022) Stock forest model based on random forest. In: Huang X, Zhang F (eds) Economic and business management. CRP Press, The Netherlands

Li Y, Gong M, Zhang XY, Koh L (2018) The impact of environmental, social, and governance disclosure on firm value: the role of CEO power. Br Account Rev 50(1):60–75. https://doi.org/10.1016/j.bar.2017.09.007

Lintner J (1965) The valuation of risk assets and the selection of risky investments in stock portfolios and capital budgets. Rev Econ Stat 47(1):13–37. https://doi.org/10.2307/1924119

Liu W (2006) A liquidity-augmented capital asset pricing model. J Financ Econ 82(3):631–671. https://doi.org/10.1016/j.jfineco.2005.10.001

Loughran T, Wellman JW (2011) New evidence on the relation between the enterprise multiple and average stock returns. J Financ Quant Anal 46(6):1629–1650. https://doi.org/10.1017/S0022109011000445

Markowitz HM (1952) Portfolio selection. J Finance 7:77–91

Marozva G (2019) Liquidity and stock returns: new evidence from Johannesburg Stock Exchange. J Developing Areas 53(2):79–90

Matemilola BT, Bany-Ariffin AN, Azman-Saini WNW (2013) Impact of leverage and managerial skills on shareholders' return. Procedia Econ Finance 7:103–115. https://doi.org/10.1016/S2212-5671(13)00224-4

Mirtaheri SL, Shahbazian R (2022) Machine learning: theory to application. CRC Press, Boca Raton

Nadyayani DAD, Suarjaya AAG (2021) The effect of profitability on stock return. Am J Humanit Soc Sci Res (AJHSSR) 5(1):695–703

Onan A (2018) A clustering based classifier ensemble approach to corporate bankruptcy prediction. J Oper Res, Stat, Econometrics Manag Inf Syst 6(2):365–376. https://doi.org/10.17093/alphanumeric.333785

Papageorgiou D, Doumpos M, Zopounidis C, Pardolos PM (2008) Credit rating systems: regulatory framework and comparative evaluation of existing methods. In: Zopounidis C, Doumpos M, Pardolos PM (eds) Handbook of financial engineering. Springer Science+Business Media, USA, pp 457–488

Pástor Ľ, Stambaugh RF (2003) Liquidity risk and expected stock returns. J Polit Econ 111(3):642–685. https://doi.org/10.1086/374184

Prathom K, Sujitapan C (2024) Performance of logistic regression and support vector machine conjunction with the GIS and RS in the landslide susceptibility assessment: case study in Nakhon Si Thammarat, southern Thailand. J King Saud Univ—Sci 36(8). https://doi.org/10.1016/j.jksus.2024.103306

Qureshi MA, Kirkerud S, Theresa K, Ahsan T (2020) The impact of sustainability (environmental, social, and governance) disclosure and board diversity on firm value: the moderating role of industry sensitivity. Bus Strateg Environ 29(3):1199–1214. https://doi.org/10.1002/bse.2427

Schonlau M, Zou RY (2020) The random forest algorithm for statistical learning. Stand Genomic Sci 20:3–29. https://doi.org/10.1177/1536867X20909688

Seok J, Kim Y, Oh YK (2024) How ESG shapes firm value: the mediating role of customer satisfaction. Technol Forecast Soc Chang 208:123714. https://doi.org/10.1016/j.techfore.2024.123714

Sharpe WF (1964) Capital asset prices: a theory of market equilibrium under conditions of risk. J Finance 19(3):425–442. https://doi.org/10.1111/j.1540-6261.1964.tb02865.x

Sumathi S, Rajappa SV, Kumar LA, Paneerselvam (2022) Machine learning for decision sciences with case studies in python. CRS Press, Boca Raton

Tan Y, Zhu Z (2022) The effect of ESG rating events on corporate green innovation in China: the mediating role of financial constraints and managers' environmental awareness. Technol Soc 68:101906. https://doi.org/10.1016/j.techsoc.2022.101906

Xie J, Nozawa W, Yagi M (2019) Do environmental, social, and governance activities improve corporate financial performance? Bus Strateg Environ 28(2):286–300. https://doi.org/10.1002/bse.2224

Yi Y, Demirel P (2023) The impact of sustainability-oriented dynamic capabilities on firm growth: investigating the green supply chain management and green political capabilities. Bus Strateg Environ 32(8):5873–5888. https://doi.org/10.1002/bse.3453

Zhang H, Shi Y, Yang X, Zhou R (2021) A firefly algorithm modified support vector machine for the credit risk assessment of supply chain finance. Res Int Bus Finance 58. https://doi.org/10.1016/j.ribaf.2021.101482

Zheng Q, Wang L, He J, Li T (2023) KNN-based consensus algorithm for better service level agreement in blockchain as a service (BaaS) systems. Electronics 12(6):1429. https://doi.org/10.3390/electronics12061429

Predicting Borsa Istanbul Banking and Finance Stocks Using Turkish Social Media Sentiment with Machine and Deep Learning

Deniz Sevinç

Abstract This study examines the impact of social media sentiment on stock prices in the banking and finance sector of Turkey, focusing on stocks that have been detected to have been manipulated by the Capital Markets Board (CMB). Using comments from Twitter and Investing.com, the study analyzes both general sentiment trends and specific suspicious actions associated with stocks. Using a binary sentiment classification approach, the study distinguishes between (1) three classification models (positive, neutral, negative) and (2) a binary classification that identifies suspicious and neutral comments. Advanced machine learning and deep learning models are used in the study to generate predictive insights. Testing various feature combinations, the study finds that the most successful models are LSTM and CNN, achieving the highest predictive accuracy when Twitter sentiment is combined with the suspicious score, and outperforming traditional models based solely on historical price trends and Investing.com sentiment data. The results demonstrate the strong predictive potential of Twitter and highlight the role of social media in indicating and potentially driving suspicious market behavior. This study highlights the need to monitor sentiment trends in markets, providing practical insights for investors and regulators.

Keywords Sentiment analysis · Twitter · Investing.com · Machine learning · Deep learning · Price prediction

1 Introduction

In recent years, the internet and social media have emerged as a powerful force in financial markets, where the rapid transmission of information has begun to rapidly influence investor sentiment and market prices. Twitter, Reddit, YouTube, Telegram, and specialized financial forums such as StockTwits and Yahoo! Finance

D. Sevinç (✉)
The Faculty of Economics and Administrative Sciences, Anadolu University, Eskişehir, Türkiye
e-mail: denizsevinc@anadolu.edu.tr

M. Gün and B. Kartal (eds.), *Machine Learning in Finance*, Contributions to Finance and Accounting, https://doi.org/10.1007/978-3-031-83266-6_9

have become popular among investors due to their ability to provide real-time news, commentary, and speculation. This trend has transformed information consumption in the financial system, indicating that social media, rather than traditional media, has now become an important tool for understanding market dynamics. Incidents such as the GameStop short squeeze and high-profile lawsuits in Turkey involving companies such as SASA, HEKTS, and various IPOs demonstrate how sharing on these platforms can lead to significant and unexpected price movements beyond traditional market expectations.

Traditional financial models, which rely heavily on quantitative data from financial statements and historical stock prices, now face limitations in predicting market direction driven by real-time discussions and investor commentary. Machine learning (ML) and deep learning (DL) approaches have begun to analyze these sentiment-driven investor comments and examine how they affect prices to shape financial forecasts more effectively.

The sentiment analysis literature initially focused on the sentiment of news. Later, pioneering studies by Antweiler and Frank (2004) and Bollen et al. (2011) highlighted the greater influence of public sentiment on stock prices, particularly through platforms such as Yahoo! Finance and Twitter. As sentiment analysis tools have evolved, Twitter has emerged as a particularly valuable resource due to its ability to reflect real-time public sentiment. The study by Bollen et al. (2011) shows that herd-like sentiment patterns from Twitter can predict stock market trends. Subsequent studies have increasingly incorporated various social media sources to better understand investor sentiment and its impact on the market (Renault 2017). Nevertheless, a gap remains regarding the use of forums popular among Turkish investors—particularly sentiment data from Investing.com—and the Turkish financial market has not been sufficiently investigated in this regard. This study aims to fill this gap by analyzing sentiment from Twitter and Investing.com, and focuses specifically on how these platforms affect market dynamics in Turkey.

The study aims at two main objectives: to identify general trends in the market by conducting sentiment analysis on selected Turkish stocks, namely ADESE, ALBRK, GLRYH, and VAKFN, and to analyze suspicious posts; to test the effectiveness of ML and DL models in stock price prediction by integrating sentiment scores and skepticism indicators with traditional price data, and finally to determine whether these approaches increase the prediction accuracy. Therefore, the study uses triple classification (positive, negative, and neutral) to detect general sentiment of posts, and binary classification (suspicious and neutral) to detect suspicious posts. Finally, the study aims to determine the most effective models for sentiment-based stock price prediction by applying ML and DL models to both types of classifications in sentiment data from Twitter and Investing.com and comparing the models.

This study is divided into five main sections. Following the introduction, the second section reviews the existing literature on sentiment analysis in financial markets. The third section focuses on the sentiment analysis process, data preprocessing steps, and sentiment classification for comments collected from Twitter and Investing.com. The fourth section examines approaches that use established models, historical price data, and sentiment and skepticism scores to predict stock prices. Finally, the fifth section discusses and concludes the findings of the study.

2 Literature Review

2.1 *Emergence and Evolution of Sentiment Analysis Through Financial Markets*

The complexity of financial markets is widely recognized, with a multitude of factors, both internal and external, shaping their behavior. These factors encompass macroeconomic factors, investor behaviors, and global events (Malkiel 2003; Shostak 1997). For a long time, traditional methods in financial analysis—relying on historical data and statistical models—have been used to predict stock prices, which rely on past data and statistical models, have been employed to predict stock prices. Nevertheless, these methods frequently struggle to completely convey the complex interplay of forces that exists within financial markets (Fama 1970; Oberlechner and Hocking 2004).

As researchers began to acknowledge the value of text data, among other factors, in predicting market movements, sentiment analysis in financial markets gained popularity. Researchers such as Fung et al. (2003), Mittermayer (2004), Tetlock (2007), and Wuthrich et al. (1998) initially conducted sentiment analysis primarily for news media, attributing market movements to news sentiment. These studies questioned the Efficient Market Hypothesis (EMH) by claiming that traditional financial data alone could not adequately represent investor sentiment's impact on the market.

Over time, sentiment analysis has been extended to a variety of platforms, and researchers have discovered that each source has a distinct influence on the market. For example, Antweiler and Frank (2004) analyzed Yahoo! Finance and Raging Bull, revealing that discussions within forums had a notable influence on trading volumes and market volatility. Nevertheless, their effect on stock returns was relatively lower. Similarly, Zhang and Skiena (2010) also find that news sentiment scores correlate with stock returns on the New York Stock Exchange, finding that sentiment impacts were often instantaneous but faded the following day. However, Twitter allowed researchers to discover that the impact of Twitter posts might be more longer-than-ever post than traditional media which leads to thinking social media may have a more permanent impact on investor decisions. For example, a study by Bollen et al. (2011) highlighted the potential of NLP-based sentiment analysis as an early predictor of stock prices due to a statistically significant relationship between Twitter public sentiment and stock market movements. This finding is consistent with behavioral finance theories, which claim that investor choices are largely impacted by psychological biases and collective emotion and that investors do not make perfectly rational investment decisions (Kahneman and Tversky 1984; Shleifer 2000; Thaler 1992).

Further studies have shown that the influence of comments made on social media and other third-party online platforms is growing. Kleinnijenhuis et al. (2015) conducted a study on the BP Deepwater Horizon oil spill incident and showed that negative media reactions caused a decline in BP's stock price and social media further fuelled these negative reactions. Additionally, Renault (2017) analyzed StockTwits, a distinct financial discussion platform, and discovered that negative postings could forecast the short-term volatility of the S&P 500 ETF, whereas the influence of positive postings was comparatively muted. Guo et al. (2017) conducted an analysis of the Chinese platform Xueqiu, revealing that posts made by individual investors had a considerable impact on sector indices, particularly during periods characterized by heightened interest and sentiment clustering.

The findings of these studies indicate that sentiment derived from social media and online forums, notably Twitter, exerts a prompt influence on market dynamics and can occasionally facilitate predictive insights. By harnessing real-time investor sentiment, these platforms introduce a dynamic informational component that traditional financial models often overlook, underscoring the critical role of sentiment in comprehending and forecasting market trends.

2.2 *Machine Learning and Deep Learning Application in Financial Forecasting Through Sentiment Analysis*

Machine learning (ML) and deep learning (DL) have improved financial forecasting by leveraging complex data from a variety of sources, including news, social media, and niche financial discussion forums. These approaches have started to surpass traditional statistical and econometric models' performance by including real-time sentiment data by offering a more complete picture of market dynamics.

Early studies laid the groundwork for the use of machine learning (ML) in finance by focusing on improving feature selection to increase forecast accuracy. Wang et al. (2012), for example, showed that the utilization of genetic algorithms could enhance the efficiency of ML models in time series forecasting, leading to better outcomes, particularly when dealing with complex financial datasets. Patel et al. (2015) examined various machine learning models and emphasized that Support Vector Machines (SVM) and Artificial Neural Networks (ANN) models, in particular, outperform traditional methods under uncertain market conditions.

According to studies, ML and DL models also outperform traditional models in sentiment analysis and price prediction using sentiments. For instance, Mittermayer (2004) classified news using machine learning methods, specifically applying market reactions over a 15-minute interval, and showed that the sentiment of news has a short-term effect on stock prices. In a related study, Mittermayer and Knolmayer (2006) used Support Vector Machines (SVM) to classify news into positive, neutral, and negative sentiment classes and introduced their own NewCATS system. The study shows that stock price predictions made with sentiment scores based on high-frequency news data outperform the traditional random walk model.

Nelson et al. (2017) showed that LSTM models are very effective in cryptocurrency markets and capture rapid price fluctuations. Similarly, Fischer and Krauss (2018) showed that the LSTM model, which takes into account long-term trends, outperformed standard forecasting models for S&P 500 stocks. Vargas et al. (2018), on the other hand, employed both Convolutional Neural Networks (CNN) and Recurrent Neural Networks (RNN) to measure the sentiment of financial news. This enhanced the precision of sentiment inference while increasing the accuracy of price predictions. Zhang et al. (2018) used the CNN models to analyze sentiment in financial news articles. By examining specific patterns within unstructured text, the CNN approach enhanced predictions of stock prices.

Eachempati et al. (2021) studied the influence of COVID-19 on stock prices using Twitter sentiment, revealing a direct relationship between the sentiment of tweets and market volatility. Zhang et al. (2021) presented a sentiment-driven adversarial learning approach combining Conditional Generative Adversarial Networks (CGAN) and Long Short-Term Memory (LSTM) models, and they evaluated the capacity of Twitter sentiment to predict stock prices and enhanced forecast accuracy with this approach. Their results showed that Twitter sentiment could be a valuable tool in predicting stock prices, especially when combined with advanced machine learning techniques.

Numerous research exist on sentiment analysis and price predictions. Nonetheless, research on Turkish text-based data is still in progress. Bozma and Kul (2020) conducted a study in which they gathered tweets from three technological firms (Alcatel, Turkcell, and Vestel) and used Support Vector Machines (SVM) for the classification of these tweets. The use of a BEKK-GARCH model to evaluate the influence of tweets on stock volatility reveals that the sentiment scores of tweets about Turkcell and Vestel significantly affect Alcatel's price volatility. Additionally, Kilimci and Duvar (2020) collected posts from many platforms, including Twitter, Bigpara, KAP, and Mynet Finance, and assessed Turkish comments' influence on the stock performance of Turkish banks utilizing hybrid deep learning models. Their findings indicate that the hybrid models Word2Vec-LSTM and FastText-LSTM provide highly accurate predictions for the stock performance of major Turkish banks. Ismayil and Demir (2023) conducted sentiment analysis for tweets of two Turkish airlines listed on Borsa Istanbul and found that there is a positive relationship between the sentiment scores of these tweets and the stock performance of the airlines. Cam et al. (2024) utilized machine learning classifiers to examine Turkish tweets, demonstrating that machine learning models, particularly SVM and Multilayer Perceptrons, can accurately forecast BIST30 movements.

However, there is still a lack of research that specifically analyzes sentiment in the Turkish language or uses local platforms such as Investing.com, which are popular among Turkish investors. Given the limited focus on the Turkish market in the existing literature, this study contributes to the literature by examining localized sentiment data from both Twitter and Investing.com. Extending sentiment analysis to include comments on both platforms can more effectively reveal market dynamics specific to Turkey, helping investors and regulators identify speculative trends and make more accurate price predictions.

3 Sentiment Analysis for Twitter and Investing.com

Sentiment analysis involves determining the emotional tone behind words to understand the sentiment expressed in text data. In this study, comments from Investing.com and Twitter were used to evaluate the impact of public opinion on stock price movements. Sentiment analysis can capture market sentiment, often an indicator of future price direction.

Sentiment analysis aims to classify text data from Twitter and Investing.com into sentiment categories. This task is divided into two classification problems: (1) a three-class sentiment classification (positive, neutral, and negative), and (2) a binary classification identifying comments as either suspicious or neutral.

This analysis focuses on selected stocks from the banking and finance sectors, specifically those identified by the Capital Markets Board (CMB) of Turkey for manipulative practices. The stocks include ADESE (Adese Real Estate Investment Inc.), ALBRK (Albaraka Türk Participation Bank Inc.), GLRYH (Güler Investment Holding Inc.), and VAKFN (Vakif Financial Leasing Inc.).

3.1 Data Collection

Tweets from Twitter were gathered using the Twitter Academic Research API, spanning from January 1, 2018, to December 31, 2022. The timeframe was chosen to ensure regularity in data, and it ends in 2022, aligning with the discontinuation of the Twitter API service. User comments on stock-related forums were obtained via web scraping from Investing.com over the same period, contributing to a rich dataset for cross-comparison.

3.2 Preprocessing

The text data was preprocessed to remove noise, such as stopwords, special characters, and numbers. Lemmatization was applied afterward to standardize the words, enhancing consistency across the dataset. The cleaned text was transformed into numerical vectors using Term Frequency-Inverse Document Frequency (TF-IDF). This method captures the significance of each term within documents, making it suitable for input into models.

The TF-IDF score for a term t in a document d is calculated as:

$$TF_{IDF(t,d)} = TF(t,d)x\log\left(\frac{N}{DF(t)}\right) \quad (1)$$

where N is the total number of documents, and $DF(t)$ is the number of documents containing the term.

3.3 Model Architectures

Several models were applied for sentiment classification. Logistic Regression, a linear model, was used to estimate the probability of a particular class using the sigmoid function:

$$P(y = 1|x) = \frac{1}{1 + e^{-w^T x + b}} \tag{2}$$

Support Vector Machines (SVM) aimed to maximize the separation margin between classes by finding an optimal hyperplane, with the decision function defined as:

$$f(x) = \text{sign}\left(w^T x + b\right) \tag{3}$$

Long Short-Term Memory (LSTM) networks, a form of Recurrent Neural Network (RNN), were utilized to detect temporal patterns in textual emotion. LSTM cells manage information flow via forget, input, and output gates.

The LSTM processes input data at each time step to update its cell state C_t and hidden state h_t are as follows:

$$f_t = \sigma\left(W_f.\left[h_{t-1}, x_t\right] + b_f\right) \tag{4}$$

$$i_t = \sigma\left(W_i \cdot \left[h_{t-1}, x_t\right] + b_i\right) \tag{5}$$

$$C_t = f_t \cdot C_{t-1} + i_t \cdot \tanh\left(W_C \cdot \left[h_{t-1}, x_t\right] + b_c\right) \tag{6}$$

$$o_t = \sigma\left(W_o \cdot \left[h_{t-1}, x_t\right] + b_o\right) \tag{7}$$

$$h_t = o_t \cdot \tanh \tag{8}$$

Convolutional Neural Networks (CNNs) were also used to capture local information in textual data using convolutional filters applied to the input sequences.

The convolution operation is expressed as:

$$(x * w)(i) = \sum_k x(k)w(i - k) \tag{9}$$

Finally, RNNs were used to capture dependencies in sequences via recurrent connections. Unlike LSTMs, RNNs lack the memory cell structure, making them better suited for shorter sequences.

RNNs update their hidden state as follows:

$$h_t = \tanh(W_h \cdot h_{t-1} + W_x \cdot x_t) \tag{10}$$

The evaluation of the models was conducted through the metrics of accuracy, precision, recall, and F1-score.

Accuracy measures the overall validity of the model's predictions:

$$Accuracy = \frac{TP + TN}{TP + TN + FP + FN} \tag{11}$$

where TP is a true positive, TN is a true negative, FP is a false positive, and FN is a false negative. Precision is the proportion of correct positive predictions:

$$Precision = \frac{TP}{TP + FP} \tag{12}$$

Recall measures the proportion of actual positives that were identified:

$$Recall = \frac{TP}{TP + FN} \tag{13}$$

Finally, the F1-score is the harmonic mean of precision and recall:

$$F1{-}score = 2 \times \frac{\Pr ecision \times \text{Re}call}{Precision + Recall} \tag{14}$$

3.4 Sentiment Analysis Results

The sentiment analysis task was conducted across two classification schemes: three-class sentiment classification (positive, neutral, negative) and binary classification (suspicious vs neutral). The models were trained and evaluated on Twitter and Investing.com data with accuracy, precision, recall, and F1-score performance metrics.

Tables 1 and 2 show the performance metrics for sentiment analysis models across different datasets (Twitter and Investing.com) for both three-class (positive, neutral, negative) and binary (suspicious vs neutral) classifications. Across the datasets (ADESE, ALBRK, GLRYH, VAKFN) for Twitter and Investing.com, LSTM consistently demonstrates the highest accuracy for most cases, achieving values of around

0.86 or higher. Regarding precision, recall, and F1-score, LSTM generally outperforms other models, making it a strong candidate for capturing nuances in sentiment, especially in complex three-class scenarios. Although CNN and RNN also perform well in some cases, they tend to have slightly lower precision and recall than LSTM for three-class classification (Table 1).

Table 1 Performance metrics of sentiment analysis models for three-class: positive, neutral, negative on Twitter & Investing.com

ADESE

	Twitter					Investing.com				
Metrics	Logistic regression	SVM	LSTM	CNN	RNN	Logistic regression	SVM	LSTM	CNN	RNN
Accuracy	0.82	0.83	0.86	0.82	0.83	0.86	0.86	0.86	0.85	0.85
Precision	0.82	0.82	0.84	0.76	0.70	0.82	0.87	0.85	0.85	0.81
Recall	0.69	0.71	0.74	0.74	0.53	0.78	0.78	0.78	0.78	0.78
F1-Score	0.73	0.74	0.75	0.75	0.72	0.81	0.81	0.82	0.81	0.81

ALBRK

	Twitter					Investing.com				
Metrics	Logistic regression	SVM	LSTM	CNN	RNN	Logistic regression	SVM	LSTM	CNN	RNN
Accuracy	0.86	0.85	0.87	0.84	0.83	0.81	0.80	0.83	0.78	0.80
Precision	0.85	0.84	0.86	0.83	0.82	0.8	0.79	0.82	0.77	0.79
Recall	0.85	0.83	0.85	0.84	0.82	0.79	0.78	0.81	0.78	0.80
F1-Score	0.85	0.84	0.86	0.84	0.82	0.79	0.79	0.82	0.78	0.79

GLRYH

	Twitter					Investing.com				
Metrics	Logistic regression	SVM	LSTM	CNN	RNN	Logistic regression	SVM	LSTM	CNN	RNN
Accuracy	0.82	0.83	0.86	0.82	0.83	0.86	0.86	0.86	0.85	0.85
Precision	0.82	0.82	0.84	0.76	0.7	0.82	0.87	0.85	0.85	0.81
Recall	0.69	0.71	0.74	0.53	0.78	0.78	0.78	0.79	0.78	0.78
F1-Score	0.73	0.74	0.75	0.72	0.81	0.81	0.81	0.82	0.81	0.81

VAKFN

	Twitter					Investing.com				
Metrics	Logistic regression	SVM	LSTM	CNN	RNN	Logistic regression	SVM	LSTM	CNN	RNN
Accuracy	0.82	0.83	0.86	0.82	0.83	0.86	0.86	0.86	0.85	0.85
Precision	0.82	0.82	0.84	0.76	0.70	0.82	0.87	0.85	0.85	0.81
Recall	0.69	0.71	0.74	0.53	0.78	0.78	0.78	0.78	0.78	0.78
F1-Score	0.73	0.74	0.75	0.72	0.81	0.81	0.81	0.82	0.81	0.81

Table 2 Performance metrics of sentiment analysis models for binary: suspicious versus neutral on Twitter and investing.com

ADESE

	Twitter					Investing.com				
Metrics	Logistic regression	SVM	LSTM	CNN	RNN	Logistic regression	SVM	LSTM	CNN	RNN
Accuracy	0.87	0.87	0.89	0.79	0.85	0.88	0.87	0.88	0.88	0.88
Precision	0.79	0.78	0.82	0.76	0.76	0.81	0.81	0.82	0.81	0.81
Recall	0.72	0.78	0.80	0.76	0.78	0.84	0.85	0.86	0.82	0.82
F1-Score	0.78	0.82	0.79	0.81	0.76	0.82	0.83	0.83	0.82	0.82

ALBRK

	Twitter					Investing.com				
Metrics	Logistic regression	SVM	LSTM	CNN	RNN	Logistic regression	SVM	LSTM	CNN	RNN
Accuracy	0.89	0.88	0.90	0.87	0.86	0.84	0.83	0.85	0.82	0.83
Precision	0.88	0.87	0.89	0.86	0.85	0.83	0.82	0.84	0.81	0.82
Recall	0.88	0.86	0.89	0.87	0.85	0.82	0.81	0.83	0.80	0.81
F1-Score	0.88	0.87	0.89	0.87	0.85	0.83	0.82	0.84	0.81	0.82

GLRYH

	Twitter					Investing.com				
Metrics	Logistic regression	SVM	LSTM	CNN	RNN	Logistic regression	SVM	LSTM	CNN	RNN
Accuracy	0.89	0.90	0.91	0.90	0.89	0.92	0.93	0.93	0.91	0.90
Precision	0.88	0.98	0.91	0.88	0.89	0.90	0.90	0.92	0.89	0.91
Recall	0.87	0.87	0.92	0.87	0.88	0.88	0.91	0.92	0.91	0.91
F1-Score	0.87	0.89	0.91	0.88	0.88	0.89	0.92	0.92	0.90	0.90

VAKFN

	Twitter					Investing.com				
Metrics	Logistic regression	SVM	LSTM	CNN	RNN	Logistic regression	SVM	LSTM	CNN	RNN
Accuracy	0.89	0.90	0.91	0.90	0.89	0.91	0.93	0.93	0.91	0.90
Precision	0.86	0.90	0.91	0.88	0.89	0.91	0.93	0.92	0.92	0.91
Recall	0.87	0.89	0.92	0.89	0.88	0.90	0.90	0.92	0.91	0.89
F1-Score	0.86	0.89	0.91	0.88	0.88	0.90	0.92	0.92	0.91	0.90

For binary classification (suspicious vs neutral), LSTM and SVM outperform in accuracy, precision, recall, and F1-score, particularly for the GLRYH and VAKFN (Table 2). LSTM achieves accuracy values of around 0.91 or higher in most cases, which indicates its robustness in differentiating between suspicious and neutral sentiments. LSTM consistently demonstrated the highest accuracy, precision, recall, and F1 score across both classification types and datasets, indicating its strength in handling diverse sentiment analysis scenarios. Given these results, LSTM is recommended as the primary model for sentiment analysis, particularly when detailed and nuanced sentiment differentiation is required.

The insights obtained from the sentiment analysis are critical for the subsequent price prediction analysis. Given LSTM's strong performance in capturing sentiment nuances, its output is expected to enhance the accuracy of predicting future market movements. Therefore, the next step in this research should involve integrating the sentiment scores derived from the LSTM model into the price prediction framework, which will likely provide valuable predictive power regarding market behavior.

4 Stock Price Forecasting: Model Design and Performance Analysis

The historical price data was collected for selected stocks from Borsa Istanbul, which were identified as subject to manipulative activities by the CMB of Turkey. The stock price data spanned from January 1, 2018, to December 31, 2022, focusing on closing prices for each trading day. For each stock, seven combinations of features were tested to predict the next day's closing price:

1. Previous day's closing price: $X_1 = [P_{t-1}]$
2. Previous day's closing price + Investing.com sentiment score: $X_2 = [P_{t-1}, S_{t-1}]$
3. Previous day's closing price + Investing.com suspicious score: $X_3 = [P_{t-1}, Q_{t-1}]$
4. Previous day's closing price + Investing.com sentiment and suspicious scores: $X_4 = [P_{t-1}, S_{t-1}, Q_{t-1}]$
5. Previous day's closing price + Twitter sentiment score: $X_5 = [P_{t-1}, T_{S_{t-1}}]$
6. Previous day's closing price + Twitter suspicious score: $X_6 = [P_{t-1}, T_{Q_{t-1}}]$
7. Previous day's closing price + Twitter sentiment and suspicious scores: $X_7 = [P_{t-1}, T_{S_{t-1}}, T_{Q_{t-1}}]$

where:

P_{t-1} represents the previous day's closing price.
S_{t-1} is the Investing.com sentiment score.
Q_{t-1} is Investing.com suspicious score.
$T_{S_{t-1}}$ is the Twitter sentiment score.
$T_{Q_{t-1}}$ is the Twitter suspicious score.
The target variable y_t for all combinations is the stock price on day t.

4.1 Models Implemented

Five different models were employed for the price prediction task. Each model was trained and evaluated on the seven feature combinations described above.

Linear regression assumes a linear relationship between the input features and the target variable. The model can be expressed as:

$$y_t = \beta_0 + \beta_1 X_{t-1} + \varepsilon \tag{15}$$

where β_0 is the intercept, β_1 is the vector of feature coefficients, and ε is the error term.

SVM for regression (SVR) aims to minimize the error within a margin by solving the optimization problem:

$\min_{w,t} w^2$ subject to

$$\left|y_t - \left(w^T X_t + b\right)\right| \leq \varepsilon \tag{16}$$

where w and b define the hyperplane, and ε is the tolerance margin for the error.

LSTM is designed to capture long-term dependencies in time series. The state update equations for the LSTM cells are given by:

$$f_t = \sigma\left(W_f.\left[h_{t-1}, X_t\right] + b_f\right) \tag{17}$$

$$i_t = \sigma\left(W_i \cdot \left[h_{t-1}, X_t\right] + b_i\right) \tag{18}$$

$$C_t = f_t \cdot C_{t-1} + i_t \cdot \tanh\left(W_C \cdot \left[h_{t-1}, X_t\right] + b_c\right) \tag{19}$$

$$o_t = \sigma\left(W_o \cdot \left[h_{t-1}, X_t\right] + b_o\right) \tag{20}$$

$$h_t = o_t \cdot \tanh \tag{21}$$

CNNs were applied to capture patterns in the sequential stock price data. The convolution operation is defined as:

$$y_t = \sum_{k=1}^{K} w_k \cdot X_{t-k} \tag{22}$$

where w_k are the convolution filters, and K is the filter size.

Finally, RNNs model temporal dependencies by maintaining hidden states that evolve over time. The update rule for the RNN hidden state is:

$$h_t = \tanh(W_h X_t + U_h h_{t-1} + b_h) \tag{23}$$

where W_h, U_h, and b_h are the weights and biases, and h_t is the hidden state at time t.

Price forecasting models were evaluated using three key metrics: Mean Absolute Error (MAE), Root Mean Squared Error (RMSE), and Mean Absolute Percentage Error (MAPE). These metrics are widely used in regression tasks to measure the accuracy of predictions against actual values.

Mean Absolute Error (MAE) measures the average magnitude of the errors in a set of predictions, without considering their direction. It is calculated as:

$$MAE = \frac{1}{n}\sum_{i=1}^{n}|y_i - \hat{y}_i| \tag{24}$$

where y_i represents the actual value, $\hat{y}_i$ indicates the predicted value, and n represents the total number of observations.

The Root Mean Square Error (RMSE) quantifies the square root of the mean of the squared discrepancies between predicted and observed values. This metric assigns greater significance to larger errors, thereby rendering it particularly responsive to outliers. It is computed as:

$$RMSE = \sqrt{\frac{1}{n}\sum_{i=1}^{n}(y_i - \hat{y}_i)^2} \tag{25}$$

Mean Absolute Percentage Error (MAPE) expresses prediction error as a percentage, making it easier to assess the error about actual values. It is defined as:

$$MAPE = \frac{100}{n}\sum_{i=1}^{n}\left|\frac{y_i - \hat{y}_i}{y_i}\right| \tag{26}$$

4.2 Price Forecasting Results

In this part of the study, price prediction models for ADESE, ALBRK, GLRYH, and VAKFN stocks are constructed with various combinations of historical stock prices and Twitter and Investing.com sentiment scores obtained from sentiment analysis in the previous section. Linear Regression, SVR, LSTM, CNN, and RNN were used as price predictions, and the performance of these models was measured and compared by MAE, RMSE, and MAPE metrics for each combination as shown in Tables 3 and 4.

In an analysis of the predictive capabilities of sentiment scores derived from Twitter and Investing.com across various models and stocks, it is observed that sentiment scores from Twitter generally exhibit marginally superior performance compared to those from Investing.com, especially when integrated with price data.

Table 3 Performance metrics for stock price prediction across models and feature combinations for ADESE and ALBRK

ADESE

Model	Performance metric	X_1	X_2	X_3	X_4	X_5	X_6	X_7
Linear regression	MAE	0.142	0.142	0.142	0.140	0.142	0.142	0.140
	RMSE	0.345	0.345	0.345	0.340	0.345	0.345	0.340
	MAPE	5.860	5.860	5.870	5.650	5.860	5.860	5.650
SVR	MAE	0.141	0.143	0.140	0.138	0.143	0.142	0.138
	RMSE	0.342	0.342	0.342	0.338	0.342	0.342	0.338
	MAPE	5.761	5.762	5.761	5.623	5.761	5.760	5.620
LSTM	MAE	0.223	0.319	0.227	0.255	0.245	0.235	0.239
	RMSE	0.386	0.484	0.397	0.451	0.454	0.431	0.441
	MAPE	5.920	5.610	5.390	5.450	5.370	5.680	5.460
CNN	MAE	0.158	0.151	0.146	0.145	0.139	0.145	0.144
	RMSE	0.372	0.370	0.357	0.340	0.347	0.348	0.342
	MAPE	5.980	5.640	5.710	6.040	5.490	5.870	5.810
RNN	MAE	0.190	0.162	0.148	0.146	0.146	0.143	0.151
	RMSE	0.376	0.343	0.353	0.342	0.358	0.347	0.379
	MAPE	5.161	7.563	6.282	6.134	5.711	6.010	5.853

ALBRK

Model	Performance metric	X_1	X_2	X_3	X_4	X_5	X_6	X_7
Linear regression	MAE	0.049	0.049	0.049	0.049	0.049	0.049	0.049
	RMSE	0.081	0.081	0.081	0.081	0.081	0.081	0.081
	MAPE	2.960	2.960	2.960	2.960	2.960	2.960	2.960
SVR	MAE	0.049	0.049	0.049	0.050	0.049	0.049	0.050
	RMSE	0.081	0.081	0.081	0.082	0.081	0.081	0.082
	MAPE	2.940	3.030	2.940	3.030	2.940	3.030	3.030
LSTM	MAE	0.068	0.107	0.148	0.113	0.140	0.079	0.115
	RMSE	0.097	0.143	0.113	0.159	0.179	0.114	0.157
	MAPE	4.370	7.010	5.660	3.180	9.180	3.700	4.720
CNN	MAE	0.067	0.054	0.059	0.056	0.056	0.054	0.058
	RMSE	0.103	0.089	0.089	0.087	0.089	0.088	0.087
	MAPE	4.110	3.350	3.770	3.490	3.580	3.350	3.350
RNN	MAE	0.108	0.052	0.067	0.059	0.066	0.061	0.057
	RMSE	0.152	0.067	0.078	0.085	0.085	0.086	0.085
	MAPE	6.950	3.230	3.880	3.650	3.530	3.950	3.300

Table 4 Performance metrics for stock price prediction across models and feature combinations for GLRYH and VAKFN

GLRYH

Model	Performance metric	X_1	X_2	X_3	X_4	X_5	X_6	X_7
Linear regression	MAE	0.159	0.159	0.159	0.158	0.159	0.159	0.158
	RMSE	0.459	0.459	0.459	0.458	0.459	0.459	0.458
	MAPE	4.510	4.520	4.500	4.450	4.520	4.500	4.450
SVR	MAE	0.164	0.164	0.164	0.162	0.164	0.164	0.162
	RMSE	0.460	0.460	0.460	0.458	0.460	0.460	0.458
	MAPE	4.670	4.660	4.700	4.680	4.660	4.700	4.680
LSTM	MAE	0.253	0.309	0.229	0.258	0.317	0.308	0.312
	RMSE	0.573	0.552	0.488	0.569	0.535	0.517	0.575
	MAPE	8.360	7.710	8.840	8.400	7.260	7.420	7.020
CNN	MAE	0.168	0.178	0.174	0.170	0.168	0.172	0.171
	RMSE	0.465	0.460	0.461	0.460	0.460	0.461	0.471
	MAPE	4.540	5.520	5.350	5.350	5.080	5.500	8.010
RNN	MAE	0.350	0.166	0.173	0.195	0.210	0.170	0.186
	RMSE	0.576	0.460	0.468	0.468	0.482	0.460	0.467
	MAPE	6.440	4.910	4.680	6.790	6.500	5.120	6.400

VAKFN

Model	Performance metric	X_1	X_2	X_3	X_4	X_5	X_6	X_7
Linear regression	MAE	0.061	0.063	0.061	0.066	0.063	0.061	0.066
	RMSE	0.143	0.142	0.143	0.143	0.142	0.143	0.143
	MAPE	4.960	5.640	4.920	5.620	5.640	4.920	5.620
SVR	MAE	0.078	0.078	0.077	0.078	0.078	0.077	0.078
	RMSE	0.145	0.145	0.145	0.145	0.145	0.145	0.145
	MAPE	6.010	6.280	7.720	6.730	6.280	6.720	6.730
LSTM	MAE	0.247	0.130	0.225	0.260	0.195	0.188	0.228
	RMSE	0.288	0.194	0.280	0.258	0.253	0.329	0.253
	MAPE	5.350	5.850	5.120	5.570	5.210	5.450	5.460
CNN	MAE	0.059	0.054	0.065	0.059	0.065	0.059	0.080
	RMSE	0.142	0.134	0.145	0.133	0.145	0.140	0.147
	MAPE	4.000	5.900	5.330	5.450	4.990	4.900	5.290
RNN	MAE	0.065	0.083	0.095	0.078	0.085	0.080	0.086
	RMSE	0.143	0.149	0.152	0.143	0.144	0.145	0.136
	MAPE	6.660	5.710	7.890	8.620	6.460	4.920	5.230

Table 3 summarizes the results for ADESE and ALBRK, showing the performance metrics of all models across various feature combinations for both stocks. For ADESE, the performance metrics indicate that CNN and LSTM models outperform simpler models like Linear Regression and SVR. The use of sentiment data, particularly the combinations X_5 (Twitter sentiment score) and X_6 (Twitter suspicious score), improved model accuracy, as indicated by a reduction in MAPE. For instance, the CNN model achieved a lower RMSE (0.340) when using Twitter sentiment and suspicious scores together in X_7. The suspicious score, which captures potential market manipulation or anomalies, proved to be a significant feature in enhancing prediction performance. This score helped detect abnormal stock movements that are not captured by regular sentiment analysis. Figure 1a illustrates that CNN was better able to capture the small fluctuations in ADESE's stock price, particularly when sentiment and suspicious data were included.

For ALBRK, the LSTM model showed the best performance overall, especially when Twitter suspicious scores were incorporated, as shown in Table 3. The LSTM model achieved the lowest MAPE when using the feature combination X_7 (Twitter sentiment and suspicious scores). The CNN model also performed well but slightly lagged behind LSTM in terms of RMSE. Figure 1b indicates that LSTM closely tracks the stock price movements. Both LSTM and CNN models were able to capture the stock price's trends, but simpler models like SVR and Linear Regression struggled to handle the more complex fluctuations. Table 3 also shows that Twitter sentiment data provided better predictive power than Investing.com data for ALBRK.

For GLRYH, LSTM again proved to be the top-performing model. As seen in Table 4, LSTM had the lowest MAPE (5.460) with the feature combination X_7, which includes both Twitter sentiment and suspicious scores. Incorporating Twitter data across the combinations consistently reduced prediction errors.

Suspicious scores were particularly valuable in this stock, which may be prone to speculation and volatility. Figure 1c for GLRYH demonstrates that LSTM was especially effective at capturing the stock's volatile movements. While CNN also performed well, LSTM was better able to reflect daily fluctuations, particularly during rapid price changes. The results highlight Twitter's real-time value in predicting market sentiment for GLRYH.

For VAKFN, LSTM was again the best-performing model, achieving the lowest MAPE (5.340) when using the feature combination X_7 (Table 4). The CNN model followed closely, with good performance when using Twitter suspicious scores in X_6. However, SVR and RNN models showed higher RMSE values, indicating lower effectiveness in capturing price fluctuations. As with other stocks, the suspicious score for VAKFN was instrumental in capturing abnormal market activities that may influence price movements. Figure 1d shows that LSTM consistently provided the best predictions when both Twitter sentiment and suspicious scores were included. LSTM captured price movements better than other models, while SVR and RNN struggled with sharp price changes.

Across all four stocks, integrating Twitter sentiment and suspicious scores (X_7) consistently enhanced the predictive performance of models, particularly the more complex ones like LSTM and CNN. The suspicious score, in particular, added an

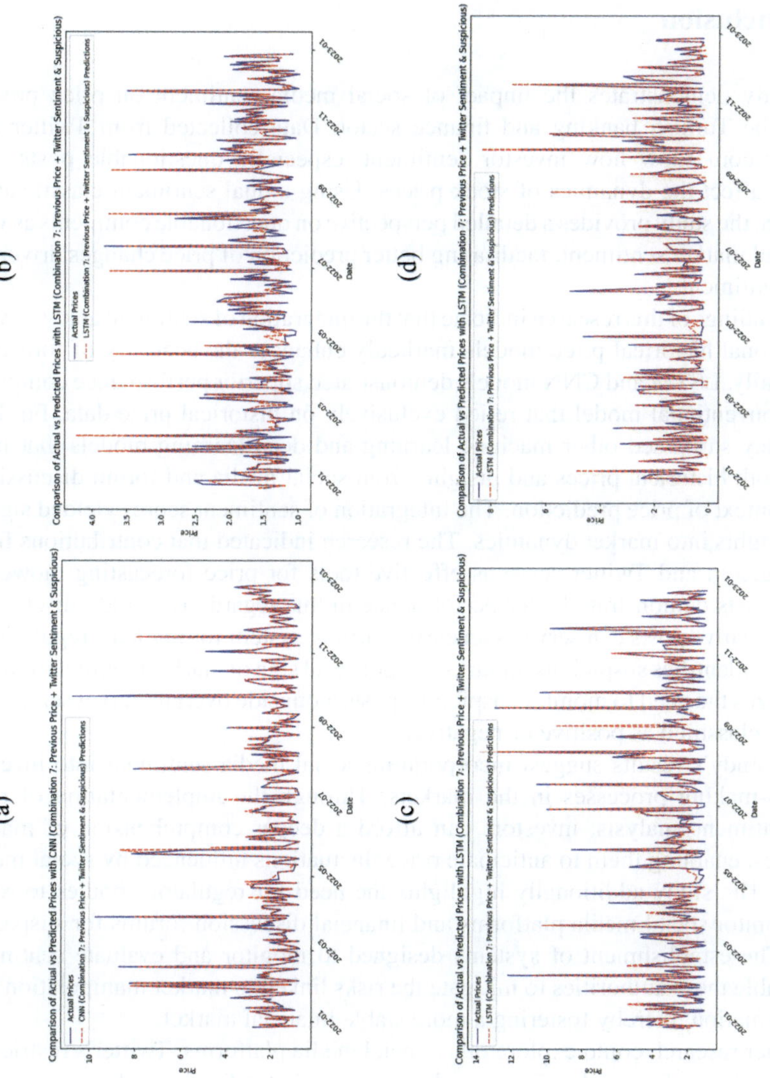

Fig. 1 Comparison of Actual versus Predicted prices for ADESE, ALBRK, GLRYH, and VAKFN

important dimension by detecting market irregularities that regular sentiment analysis might miss. These results suggest that sentiment data from social media, particularly Twitter, adds significant value to stock price prediction models.

5 Conclusion

This study demonstrates the impact of social media sentiment on price prediction in the Turkish banking and finance sector. Data collected from Twitter and Investing.com show how investor sentiment, especially questionable posts, can strongly affect the dynamics of stock prices. Using a dual sentiment classification approach, the study provides a detailed perspective on questionable comments as well as general market sentiment, facilitating better prediction of price changes driven by public sentiment.

The findings of this research indicate that the integration of sentiment analysis with conventional historical price models markedly enhances the accuracy of forecasts. Specifically, LSTM and CNN models demonstrated superior performance compared to the conventional model that relied exclusively on historical price data. Furthermore, they surpassed other machine learning and deep learning models that integrated both historical prices and insights from social media and forum discussions in the context of price prediction. The integration of sentiment scores yielded significant insights into market dynamics. The research indicated that contributions from Investing.com and Twitter serve as effective tools for price forecasting; however, Twitter posts demonstrate better performance in this regard. The study discovered that speculative posts can serve as an early warning system for market irregularities. The study identifies suspicious remarks as a potential sign of market manipulation and emphasizes the need to monitor suspicious posts alongside overall market sentiment, which is classified as positive or negative.

This study's results suggest incorporating social media sentiment into investor decision-making processes in the markets. Through the implementation of real-time sentiment analysis, investors can afford a deeper comprehension of market dynamics, enabling them to anticipate price fluctuations influenced by social media activity. The study additionally highlights the need for regulatory bodies to regularly monitor social media platforms and financial discussion forums for suspected fraud. The establishment of systems designed to monitor and evaluate sentiment may enable these authorities to mitigate the risks linked to market manipulation and misinformation, thereby fostering a more stable financial market.

Further research could explore other social media platforms. Twitter's restriction of its Academic Research API in particular necessitates future studies to examine alternative platforms. Furthermore, expanding the study to other markets and sectors would further validate the applicability of these results.

References

Antweiler W, Frank MZ (2004) Is all that talk just noise? The information content of internet stock message boards. J Financ 59(3):1259–1294. https://doi.org/10.1111/j.1540-6261.2004.00662.x

Bollen J, Mao H, Zeng X (2011) Twitter's mood predicts the stock market. J Comput Sci 2(1):1–8. https://doi.org/10.1016/j.jocs.2010.12.007

Bozma G, Kul S (2020) Twitter ile Hisse Senetleri Oynaklığı Tahmin Edilebilir mi? Sosyoekonomi 28(45):315–326. https://doi.org/10.17233/sosyoekonomi.2020.03.18

Cam H, Cam AV, Demirel U, Ahmed S (2024) Sentiment analysis of financial Twitter posts on Twitter with the machine learning classifiers. Heliyon 10(1):e23784. https://doi.org/10.1016/J.HELIYON.2023.E23784

Eachempati P, Srivastava PR, Panigrahi PK (2021) Sentiment analysis of COVID-19 pandemic on the stock market. Am Bus Rev 24(1):8. https://doi.org/10.37625/abr.24.1.141-165

Fama EF (1970) Efficient capital markets: a review of theory and empirical work. J Financ 25(2):417. https://doi.org/10.2307/2325486

Fischer T, Krauss C (2018) Deep learning with long short-term memory networks for financial market predictions. Eur J Oper Res 270(2):654–669. https://doi.org/10.1016/J.EJOR.2017.11.054

Fung GPC, Xu Yu J, Lam W (2003) Stock prediction: integrating text mining approach using real-time news. IEEE/IAFE conference on computational intelligence for financial engineering, proceedings (CIFEr), pp 395–402. https://doi.org/10.1109/CIFER.2003.1196287

Guo K, Sun Y, Qian X (2017) Can investor sentiment be used to predict the stock price? Dynamic analysis based on China stock market. Physica A 469:390–396. https://doi.org/10.1016/J.PHYSA.2016.11.114

Ismayil J, Demir O (2023) Relationship between Twitter activity and stock performance: evidence from Turkish airline industry. Foresight 25(5):701–715. https://doi.org/10.1108/FS-11-2021-0224/FULL/PDF

Kahneman D, Tversky A (1984) Choices, values, and frames. Am Psychol 39(4):341–350. https://doi.org/10.1037/0003-066X.39.4.341

Kilimci ZH, Duvar R (2020) An efficient word embedding and deep learning based model to forecast the direction of stock exchange market using twitter and financial news sites: a case of istanbul stock exchange (BIST 100). IEEE Access 8:188186–188198. https://doi.org/10.1109/ACCESS.2020.3029860

Kleinnijenhuis J, Schultz F, Utz S, Oegema D (2015) The Mediating role of the news in the BP oil spill crisis 2010: how U.S. news is influenced by public relations and in turn influences public awareness, foreign news, and the share price. Communication Research 42(3):408–428. https://doi.org/10.1177/0093650213510940

Malkiel BG (2003) The efficient market hypothesis and its critics. J Econ Perspect 17(1):59–82. https://doi.org/10.1257/089533003321164958

Mittermayer MA (2004) Forecasting intraday stock price trends with text mining techniques. Int Conf Data Min 37:1029–1038. https://doi.org/10.1109/HICSS.2004.1265201

Mittermayer MA, Knolmayer GF (2006) NewsCATS: a news categorization and trading system. IEEE Int Conf Data Min ICDM 1002–1007. https://doi.org/10.1109/ICDM.2006.115

Nelson DMQ, Pereira ACM, De Oliveira RA (2017) Stock market's price movement prediction with LSTM neural networks. In: Proceedings of the international joint conference on neural networks, pp 1419–1426 https://doi.org/10.1109/IJCNN.2017.7966019

Oberlechner T, Hocking S (2004) Information sources, news, and rumors in financial markets: insights into the foreign exchange market. J Econ Psychol 25(3):407–424. https://doi.org/10.1016/S0167-4870(02)00189-7

Patel J, Shah S, Thakkar P, Kotecha K (2015) Predicting stock and stock price index movement using trend deterministic data preparation and machine learning techniques. Expert Syst Appl 42(1):259–268. https://doi.org/10.1016/J.ESWA.2014.07.040

Renault T (2017) Intraday online investor sentiment and return patterns in the U.S. stock market. J Bank Finance 84:25–40. https://doi.org/10.1016/J.JBANKFIN.2017.07.002
Shleifer A (2000) Inefficient markets: an introduction to behavioral finance. Oxford University Press, In Inefficient Markets. https://doi.org/10.1093/0198292279.001.0001
Shostak F (1997) In defense of fundamental analysis: a critique of the efficient market hypothesis. Rev Austrian Econ 10(2):27–45. https://doi.org/10.1007/BF02538483
Tetlock PC (2007) Giving content to investor sentiment: the role of media in the stock market. J Financ 62(3):1139–1168. https://doi.org/10.1111/J.1540-6261.2007.01232.X
Thaler RH (1992) The Winner's Curse: Paradoxes and Anomalies of Economic Life. The Free Press. https://books.google.com.tr/books?hl=tr&lr=&id=VmDp2BpI23UC&oi=fnd&pg=PT14&dq=The+winners+curse:+Paradoxes+and+anomalies+of+economic+life&ots=doMwzcpZOd&sig=9TS-_9aM6GwJ3Ihfymmo9XsmDAg&redir_esc=y#v=onepage&q=The%20winners%20curse%3A%20Paradoxes%20and%20anomalies%20of%20economic%20life&f=false
Vargas MR, Dos Anjos CEM, Bichara GLG, Evsukoff AG (2018) Deep learning for stock market prediction using technical indicators and financial news articles. In: Proceedings of the international joint conference on neural networks https://doi.org/10.1109/IJCNN.2018.8489208
Wang JJ, Wang JZ, Zhang ZG, Guo SP (2012) Stock index forecasting based on a hybrid model. Omega 40(6):758–766. https://doi.org/10.1016/j.omega.2011.07.008
Wuthrich B, Cho V, Leung S, Permunetilleke D, Sankaran K, Zhang J, La W (1998) Daily stock market forecast from textual Web data. IEEE Int Conf Syst Man Cybern 3:2720–2725. https://doi.org/10.1109/ICSMC.1998.725072
Zhang W, Skiena S (2010) Trading strategies to exploit blog and news sentiment. Proc Int AAAI Conf Web Soc Media 4(1):375–378. https://doi.org/10.1609/icwsm.v4i1.14075
Zhang Y, Li J, Wang H, Choi SCT (2021) Sentiment-guided adversarial learning for stock price prediction. Front Appl Math Stat 7:601105. https://doi.org/10.3389/FAMS.2021.601105/BIBTEX
Zhang Z, Zhang Y, Shen D, Zhang W (2018) The dynamic cross-correlations between mass media news, new media news, and stock returns. Complexity 2018:1–11. https://doi.org/10.1155/2018/7619494

Portfolio Management Through Algorithmic Trading

Ahmet Akusta

Abstract This chapter stresses how algorithmic trading has transformed portfolio management, underscoring the resultant ability to optimize risk-adjusted returns, enhance decision processes, and sustain efficient asset allocation. Advanced computational methods utilized in this research examine algorithmic strategies as a means to address the complexities of today's financial markets in their ability to handle risk management, diversification, and periodic rebalancing. The results of the optimization of a portfolio consisting of six Nasdaq 100 stocks—Amazon, Apple, AMD, Tesla, Google, and NVIDIA—for ten years, from 2014 to 2024, are shown here. These assets have been selected based on their historical performance and variable risk-return profile as a sample to evaluate algorithmic trading strategies. In this paper, SLSQP is used to optimize the weights of each portfolio according to the Sharpe ratio, with efficient capital allocation considering the realistic constraint of no short-selling on the historical price data. Annual rebalancing was adopted to dampen the drifting of weights and to make the weights given at any period closer to the target weights. The performance of the portfolio is measured concerning the Nasdaq 100 through a set of key metrics: the cumulative return, the annualized return, volatility, and the Sharpe ratio. Hereby, the optimized portfolio gains an annualized return of 46.89% with a cumulative return of 4576.56% throughout the period under review. Although the portfolio demonstrated higher volatility (40.89%) in comparison to the Nasdaq 100, its Sharpe ratio of 1.12 surpassed that of the benchmark (0.90), thereby illustrating superior risk-adjusted performance. The rebalancing process effectively maintained the efficiency of the portfolio, although the concentration of risk in high-growth assets, such as NVIDIA, was brought to light. The findings highlight the inherent trade-offs between return maximization and risk management, offering valuable insights for investors, practitioners, and policymakers.

Keywords Asset allocation · Algorithmic trading · Machine learning · Portfolio diversification

---

A. Akusta (✉)
Konya Technical University, Konya, Türkiye
e-mail: aakusta@ktun.edu.tr

M. Gün and B. Kartal (eds.), *Machine Learning in Finance*, Contributions to Finance and Accounting, https://doi.org/10.1007/978-3-031-83266-6_10

1 Introduction

Portfolio management is the cornerstone of modern finance, whereby assets are allocated and optimized for the achievement of certain investment objectives. Portfolio management, based on Markowitz's MPT, focuses on balancing risk and return, establishing diversification boundaries, and optimizing asset allocation efficiency. With the increasing complexity of financial markets, algorithmic trading emerged as a revolutionary innovation, requiring systematic and data-driven portfolio management. Algorithmic trading utilizes computational power to improve asset allocation efficiency. It also enhances decision-making by managing large datasets with minimal human biases (Kissell 2011; Dananjayan et al. 2023).

Algorithmic trading replaced the portfolio management system based on heuristic techniques with one that relied on quantitative precision. For example, real-time TCA provides actionable insights for the optimization of trading strategies, and portfolio managers can thus minimize inefficiencies and eventually realize better risk-adjusted returns, as Kissell (2011) mentions. Indeed, this has been supportive in negotiating the volatility and complexity of modern financial markets.

ML integrated with AI in algorithmic trading has expanded its possibility in portfolio management. Deep reinforcement learning frameworks, including DeepScalper, integrate macro and micro levels of market information in the quest for optimized intraday trading strategies (Sun et al. 2021). Furthermore, algorithmic trading has proven beneficial in risk management and diversification, essential components of portfolio management. Hybrid AI techniques involving neural networks with fuzzy logic forecast market trends and manage portfolio risks concerning return objectives (Haddadian et al. 2022). Quantitative trades, such as Bollinger Bands and moving average crossovers, outperformed any traditional buy-and-hold strategy in almost all volatile or bearish markets entering 2023 (Patel 2023). These examples are illustrations of an emergent reliance on algorithmic modes of approach as a way out of the complexities of the market.

Advancements in technology have enhanced the efficiency of algorithmic trading. For example, GPU technologies increased computational efficiency in the various trading algorithms, increasing the speed with which data is worked out and the execution of trades (Topiwala and Dai 2022). Blockchain technology is being used to enhance trade transparency and security through decentralized ledgers (Goyal et al. 2024). These innovations highlight how emerging technologies form the basis for advancing robust and efficient algorithmic trading systems.

Despite such advantages, algorithmic trading has a whole set of challenges in portfolio management. Overfitting in machine learning models, regulatory constraints, and ethical issues with respect to decisions driven by AI are some obstacles identified in recent studies (Lemma 2020; Weinan et al. 2023).

This chapter takes a closer look at how algorithmic trading improves portfolio management practices. The paper discusses how algorithmic strategies increase returns, reduce risks, and maintain optimized allocations of assets through periodic portfolio rebalancing. The chapter analyzes specific key performance metrics by studying the performance of six stocks from the Nasdaq 100 index over ten years (2014–2024). It compares algorithmically optimized portfolios with traditional benchmarks in terms of return, volatility, and Sharpe ratios.

2 Literature Review

The effective integration of AT in portfolio management has completely changed today's financial markets. This literature review examines the multifaceted effects of AT on portfolio management, taking into consideration a wide array of studies exploring the influence of changes in market dynamics, trading strategies, risk management, and regulatory considerations.

Algorithmic trading has been extensively researched in terms of its impact on market liquidity, price discovery, and volatility. Zhou et al. (2020) investigate the question of whether AT would exacerbate extreme price movements on days of chaotic market conditions. Their findings indicated that, on the day of the major market decline, stocks with the most intense activity from AT were less likely to experience extreme price drops, which leads to the stipulation that AT may help dampen temporary pricing mistakes. Along these lines, Mestel et al. (2018) investigate the Austrian equity market to find that an increase in the market share of AT is associated with a reduction of quoted and effective spreads without adverse impacts on quoted depth or price impacts. The results could signal that AT enhances liquidity and fosters market efficiency through its market-making force.

However, views are far from unanimous on how the effect of AT is invariably beneficial in all markets. Ramos and Perlin (2020) investigate the equities market in Brazil. They find that evidence of AT decreasing liquidity contradicts that of more developed markets, which may indicate that due to issues of market maturity or regulatory environments, the effect of AT can vary in emerging markets. Also, Hassavayukul (2020) found that in the Thai market, AT was associated with a decline in liquidity proxies for effective half-spread and share turnover.

Muravyev and Picard (2022) analyze the effects of trading activity induced by AT on liquidity and volatility. Therein, they find that periodic surges in trading activity—a consequence of algorithms probably running in a synchronized fashion—result in increased volatility, largely unaccompanied by improvements in liquidity. This finding also brings into question the stability of those markets, with high participation by ATs during periods of unusual trading activity.

This development of complex algorithmic approaches catalyzed modern portfolio management practices. Sukma and Namahoot (2024), provide an in-depth study of some AT strategies based on technical indicators like RSI, EMA, and MACD. The study underlines the requirement to go for a multi-indicator approach in improving the interpretability and prediction accuracy of AT models; however, on a consistent basis, strategies that are tested fail to outperform market benchmarks.

Against the prevailing market conditions, Mengshetti et al. (2024) propose a new trading strategy—weapon candle strategy—considering RSI, EMA, volume-weighted average price, and MACD. An extensive backtesting study on the National Stock Exchange of India was done to see that the combined use of indicators provides a higher percentage of profitable trades compared to the use of single indicators. Their conclusions point out that such combined strategies can provide a more formidable framework in algorithmic portfolio management.

Furthermore, machine learning and artificial intelligence methods have increasingly been applied to algorithmic trading (AT). Levendovszky et al. (2019) present neural network-based algorithms for electronic trading, including feedforward neural networks in order to estimate forward conditional probability distributions. As these authors point out, this may provide the possibility of developing trading signals if large price movements are expected, which may further improve decision-making in portfolio management.

Alaminos et al. (2024) discuss deep learning and quantum methods on neural networks and genetic algorithms that use the inclusion of EGARCH models to estimate volatility in cryptocurrency markets. Based on their comparison study, advanced computation methods can enhance the performance of trading signals that might be helpful in portfolio management during such highly volatile conditions.

High-frequency trading (HFT), which is a subset of AT where trades are executed with very high speeds, has been a point of interest. The study by Courdent and McClelland (2022) researches the impact of HFT on market quality in South Africa. Whereas HFT contributes to increased liquidity, it is also associated with higher short-term volatility. Their study underlines that HFT does not have that simple of an impact on different aspects of market quality.

In the case of emerging markets, it may, however, be very different in the adoption and effects of AT compared with developed markets. In this context, Ersan and Ekinci (2016) present an investigation of AT participation in and the impact of HFT in Borsa Istanbul. They document an upward trend in AT activity. Their findings indicate that HFT activity is expected to be higher for larger orders and also follows the enhancement of trading platforms, implying that technological improvements may allow more AT participation. Mukerji et al. (2019) simulate the introduction of AT in asset markets to evaluate its effects on market quality. The results indicate that although there may be an initial benefit from increased AT by way of improved liquidity, further increases in AT may not necessarily produce benefits proportional to these increases, and great care should be taken concerning the role of AT in different market environments.

Jacob-Leal and Hanaki (2024) explore the effects of AT on behavior through experimental studies into the perceived presence of AT in the determination of price predictions and trading by human subjects. Even the expectation of AT impairs price convergence toward fundamental values, which would thus suggest that AT may have an effect on market dynamics beyond its direct trades.

Risk management is related to portfolio management, and AT adds other dimensions to the problem. In this regard, Azhmyakov et al. (2023), propose an optimal trading algorithm based on the principles of control engineering. They use a model-free realization of the so-called proportional integral derivative (PID) controller. They tried to address the management of inlet fluctuations in the financial market as a whole, but they were most relevant to new cryptocurrencies through the application of dynamically adjusted strategies according to market conditions.

Khurana et al. (2024) deal with the risks associated with long-term investments in volatile cryptocurrency markets. The authors have developed an optimized Greedy-Cost Averaging Trading (OG-CAT) framework, an alternative to a buy-and-hold strategy. Their approach leverages the high volatility and wavy price structures of cryptocurrencies to reduce drawdowns while offering better profitability; it shall prove to be a useful tool for a risk-averse portfolio manager.

Recent AT proliferation has raised discussions of its greater economic and institutional implications. Schmidt-Kessen et al. (2022) analyze the inflexibility inherent in automated contracts and algorithms; while this sort of constraint can increase welfare by enabling cooperation in algorithmic markets, it creates systemic risks. Their investigation of algorithmic trading in financial markets is a call for regulatory frameworks capable of damping these risks without stifling innovation.

Martins Pereira (2020) discusses the regime on algorithmic trading in the European Union and finds a serious lacuna therein: no proposal is made to regulate the mere execution algorithms employed by automated order routers. This would tend to heighten the risks involved in that even basic execution algorithms have been found to have caused market disruptions in past trading mishaps. This study emphasizes how broad regulatory strategies at every level of AT are needed to maintain stability in the markets.

Cryptocurrency markets have welcomed a new frontier for AT applications. Omran et al. (2023) present a multi-objective particle swarm optimization algorithm undertaking the cryptocurrency algorithmic trading of Litecoin. The algorithm will help traders find optimal trading strategies, balancing return on investment, risk measures such as the Sortino ratio, and the number of trades executed. These findings prove the efficiency of the proposed algorithm in different market conditions and can thus indicate the possibility of AT entering the specific conditions of cryptocurrency trading.

Martínez et al. (2022) discuss the applications of behavioral finance measures to AT. The authors outline an algorithmic trading system whose trading logic is rooted in accumulated COVID-19 case incidents. Similarly, backtests on European indices have provided indications of profitability for the strategy. The Martínez et al. results suggest that the utilization of investor sentiment and exogenous events, such as pandemic fear, in an AT model tends to result in favorable portfolio management consequences.

Improvement in techniques for data analysis and modeling has enhanced the power of AT in portfolio management. Yang et al. (2015) apply inverse reinforcement learning to infer traders' reward functions from observed actions. The authors model the trader behavior as a Markov decision process in order to classify and identify trading algorithms that can then be used in improving AT strategies.

Cooper et al. (2015) import quality control measures from industrial processes to construct new performance metrics for AT. Their multi-scale capability measure assesses, within one framework, control, expected tail loss, and risk-adjusted performance and, therefore, forms a more descriptive and proper evaluation of AT strategies than can be made using traditional financial measures.

While there are quite a number of benefits that come with AT, there are challenges that should equally be put into great consideration. A number of studies raise concerns that AT has the potential to contribute to market instability. Muravyev and Picard (2022) note that some periodicities in algorithmic trading may lead to higher volatility, probably as a result of predictable algorithmic behavior that can be exploited by others.

Regulatory issues are also high on the agenda when it comes to Schmidt-Kessen et al. (2022) and Martins Pereira (2020) pointed out that regulations are needed, given the advanced algorithms, even in the case of the simplest execution algorithms that carry significant risks. A balance between encouraging innovation and ensuring market integrity eludes both policymakers and market participants.

3 Methodology

The steps taken in attempting to apply algorithmic trading techniques for portfolio optimization are hereby described sequentially. They include objective formulation, data preparation, and then mathematical optimization of the implementation of building such a portfolio for its improved risk-adjusted performance.

3.1 Initial Setup

The main goal of this analysis is to determine the optimal portfolio composition that maximizes the Sharpe ratio, which therefore maximizes the risk-adjusted returns. For this problem, restrictions have been added to represent real investment scenarios:

- Full Capital Allocation: Portfolio weights should sum to 1.
- No Short selling: Weights were constrained to non-negative values

In this way, the constraints ensured that the portfolio reflected realistic investor behavior where there were no leveraged positions nor excessive exposure to any one asset.

The selected assets are the most dominant stocks in the Nasdaq 100 index: Amazon, Apple, AMD, Tesla, Google, and NVIDIA. These stocks have been selected based on their good returns and diversified risk profile historical performance measures, which are indicative that these assets are ideal to analyze for algorithmic trading strategies and portfolio optimization. For example, NVIDIA presents an amazing annualized return of about 98.05% reflective of its market dominance in graphics processing and AI technologies. In the same light, Amazon and Tesla had strong return profiles that underlined growth potential in both cases, despite higher volatility across one.

Each selected stock represents a different domain in the technology and innovation ecosystem; hence, taken together, they represent an overview of the entire industry. Amazon includes e-commerce and cloud computing for its dual role in consumer retail and enterprise solution areas. Apple excels in consumer technology and hardware, driving dominance in smartphones and wearables. AMD is integral to semiconductor design and manufacturing for computing and gaming applications. Tesla pioneered electric vehicles and clean energy, representing the seismic shift of automotive and energy markets. Google maintains the most-used search engine, with enormous sections of digital advertising and cloud services underpinning the information economy. Nvidia's leading position in graphics processing and AI acceleration provides a vision of the future for gaming, machine learning, and data-intense applications. Overall, the complementing set of stocks provides diverse yet connected insights into high-growth sectors, making it an exemplary case study about technological trends.

Historical price data from 2014 to 2024 was used with the purpose of capturing a decade of market conditions. This period has been selected since it presents growth phases as well as volatility periods, ensuring that the analysis covers both diverse market environments.

3.2 Data Preparation

Preprocessing the dataset for cleaning and transformation was done to ensure consistency and reliability for the analysis. Missing values were handled through interpolation to maintain time series continuity, and the dataset underwent screening for duplicate entries to prevent calculation inaccuracies. Dates were put into a uniform format to match standardized practices for time series data; this allows for easy indexing and temporal analysis. Further, the adjusted closing prices were extracted from the data to adjust for stock splits, dividends, and other corporate-level actions that provide the right premise for return calculations and performance metrics. This rigorous preprocessing ensured a strong starting dataset for the subsequent algorithmic trading and portfolio optimization analyses.

Daily returns for each asset were calculated using the formula:

$$R_{i,t} = \frac{P_{i,t} - P_{i,t-1}}{P_{i,t-1}} \tag{1}$$

where $P_{i,t}$ represents the adjusted closing price of asset i on day t. This produced a time series of percentage changes, serving as the basis for subsequent analyses.

To enable comparability across assets, daily returns and volatility were annualized:

Annualized Return (μ_{annual}):

$$\mu_{\text{annual}} = (1 + \mu_{\text{daily}})^{252} - 1 \tag{2}$$

Annualized Volatility (σ_{annual}):

$$\sigma_{\text{annual}} = \sigma_{\text{daily}} \times \sqrt{252} \tag{3}$$

Portfolio optimization was conducted to maximize the Sharpe ratio (Sharpe 1966):

$$\text{Sharpe Ratio} = \frac{E[R_p] - R_f}{\sigma_p} \tag{4}$$

where:

- $E[R_p]$: The portfolio's expected return.
- R_f: The risk-free rate, set at 1%.
- σ_p: The portfolio's volatility.

This approach balances return maximization with risk control, ensuring efficient capital allocation.

3.3 Numerical Methods

The SLSQP algorithm is used to solve this optimization problem. This technique is ideal when an optimization problem under nonlinear constraints needs to be solved since the weight allocation for every asset will be given out (Kraft 1988).

3.4 Asset Weight Selection

During the optimization process, the historical performance of each asset was factored in, with higher returns relative to its volatility favored. Indeed, as captured by the stock price chart, Nvidia indicated incredible growth during the period under study, having topped the lot by the end of 2015.

On the other hand, assets like Google and Apple took a stable position but did not reveal much growth, reflected by their flatter cumulative return curves with relatively stagnant price movements. Later, as shown in Table 1, these assets get negligible allocations (~0.00%) in the initial weight assignment.

Moderately weighted assets: Amazon with 14.78% and Tesla at 11.87%, respectively, illustrate an intermediate growth of stock price and CTR. These would be a diversification addition to the portfolio while ensuring positive contributions to its risk-return profile.

Table 1 Optimized asset weights

Asset	Optimal weight (%)
Amazon	14.78
Apple	~ 0.00
AMD	6.63
Tesla	11.87
Google	~ 0.00
Nvidia	66.73

The stock price movements until 31-12-2015 show that Nvidia, Amazon, and Tesla have a significant uptrend, while Apple and Google have somewhat flat graphs. Because of this, Nvidia and Amazon had a heavier weight in the portfolio since their performance was indicative of stronger growth potential.

4 Empirical Analysis

Empirical analysis estimates the out-of-sample performance of an algorithmically optimized portfolio using historical market data. The following section describes the historical simulation of the portfolio, its risk contribution analysis, and the key observations from its performance.

4.1 Risk Contribution Analysis

The total risk for the portfolio was decomposed into the contributions by individual assets, underscoring their relative impacts on the overall volatility. According to Table 2, one would expect that with both its dominating weight of 66.73% and high volatility, Nvidia was contributing 77.09% to the total risk of the portfolio. This is further reflected by the Annualized Return vs. Volatility chart, which has positioned Nvidia in the top-right corner, indicating its high return potential, coupled with high volatility. Similarly, Tesla, while it accounts for only 11.87% of the portfolio's weight, due to its high volatility further positioned to the right in the chart as compared to other assets, has added 9.29% to the overall portfolio risk.

Table 2 Risk contribution analysis

Asset	Weight (%)	Risk contribution (Absolute) (%)	Risk contribution (%)
Amazon	14.78	3.02	7.38
Tesla	11.87	3.80	9.29
Nvidia	66.73	31.53	77.09

The results emphasize the concentration risk inherent in the portfolio, driven by the significant allocation to Nvidia. While this concentration enables the portfolio to achieve high returns, it underscores the trade-off between maximizing returns and maintaining diversification.

The results suggest that while the portfolio is efficient in its risk-return allocation, continuous monitoring of risk contribution is essential to prevent over-reliance on a single asset and ensure sustained performance over time.

5 Rebalancing Strategy

Rebalancing, in general, is a crucial activity in portfolio management from the periodical readjustment of weights of assets toward their target allocations. This section explores how an annual rebalancing strategy can be implemented, its impact on the offered portfolio performance, and a risk management perspective.

Over time, changes in asset prices will lead to portfolio drift away from the optimized weights through what is known as 'weight drift'. This will create some unexpected risk exposures, which will degrade the efficiency of the portfolio. Rebalancing corrects these misalignments and restores intended diversification to the portfolio for continuity in reflecting the set risk-return objectives during optimization.

However, rebalancing has its own set of challenges. Transaction costs and tax implications can reduce net returns, especially with frequent adjustments. Accordingly, the determination of rebalancing frequency must balance the efficacy of rebalancing with its costs.

5.1 Implementation of the Rebalancing Strategy

For this analysis, an annual rebalancing frequency was chosen. This approach provides a balance between maintaining portfolio alignment and minimizing transaction costs. At each rebalancing point:

- Asset weights were recalculated based on current market values.
- Deviations from the target weights were corrected by buying or selling assets.
- The portfolio was adjusted to restore the optimized allocation.

Rebalancing prevented impacts from weight drift and precluded overexposure to high-performing, volatile assets like Nvidia. It is also a contributing factor to the portfolio maintaining a rather stable risk profile while various asset prices fluctuate.

5.2 *Evaluating the Costs of Rebalancing*

This analysis does not include transaction costs explicitly, but one should certainly be cognizant of their possible effects. For example, rebalancing too frequently may result in costs that gnaw net returns away, whereas in a market with high bid-ask spreads or portfolios with small amounts of investment, such may be the case. Further work could quantify these costs to complete the assessment of the rebalancing strategy.

6 Performance Evaluation

The performance needs to be appraised to establish how well the optimized portfolio has been undertaking its investment goals. This section looks at the key performance indicators of the portfolio and its results against the selected Nasdaq 100 benchmark, as well as the trade-offs made between return maximization and risk management.

6.1 *Key Metrics of Portfolio Performance*

The performance of the portfolio had been evaluated based on its cumulative return, annualized return, volatility, and Sharpe ratio. These measures provide a full picture of portfolio growth, its risks, and efficiency.

The cumulative return captures the total percentage growth of their portfolio over the investment period. Over the analysis period 2014–2024, the portfolio completed a state-of-the-art cumulative return of 4576.56%, showing its aptitude for much value appreciation with algorithmic optimization.

It means that the annualized return measures or quantifies the average yearly growth in the portfolio. Indeed, a 46.89% annualized return does show what strong growth potential the portfolio has, beating the returns of many more traditional investment strategies. The annualized volatility, 40.89%, is the amount of risk the portfolio has. This is higher than the benchmark but expected for a high-growth portfolio of the selected assets.

The Sharpe Ratio is one of the most applied measures when it comes to portfolio risk-adjusted return. It indicates how well the returns are generated against the level of risk taken. In this view, the Sharpe ratio of 1.12 will be considered relatively good concerning risk-adjusted performance. While this is generally regarded as a strong outcome, it is important to interpret the figure in light of broader market conditions and the specific characteristics of the portfolio.

This Sharpe Ratio level reflects that the portfolio realized a commendable balance of returns against its risk. However, caution must be exercised in assuming that past performance guarantees future outcomes, as market volatility and other exogenous factors may impact these risk-return dynamics. While the ratio underlines the relative

Table 3 Portfolio versus Nasdaq 100 Metrics

Metric	Portfolio	Nasdaq 100
Annualized return	0.468901	0.203247
Annualized volatility	0.408931	0.214660
Sharpe ratio	1.122199	0.900247

efficiency of the portfolio, it shall be read in conjunction with other measures and qualitative aspects to arrive at a complete perception of its all-round performance.

6.2 Benchmark Comparison

To put a greater perspective on this portfolio's performance, it is set against the Nasdaq 100, generally a proxy for growth-oriented investments. This gives some context toward understanding how these metrics for the portfolio either meet or fall short of a widely accepted benchmark. The portfolio outperformed the Nasdaq 100 across all measurements, but great leeway should be applied when interpreting such results since each has different investment strategies and risk profiles innate to them.

According to Table 3, at significantly higher annualized returns compared to the Nasdaq 100, it is likely that the portfolio effectively captured high-growth opportunities via algorithmic optimization. Meanwhile, higher returns accompany greater annualized volatility, reflecting concentrated exposure to the high-growth asset Nvidia. This increased volatility overwrites the usual trade-off between risk and reward in such an aggressive growth strategy.

On a risk-adjusted basis, the portfolio has had a Sharpe Ratio of 1.12 versus 0.90 for the Nasdaq 100. This would then indicate that there is further efficiency in the balance of return and risk. In this regard, such a comparison does indeed underline strong performance for the portfolio, and it is best taken in conjunction with qualitative factors and market conditions.

6.3 Trade-Off Analysis: Balancing Return and Risk

These superior returns of the portfolio did not come for free; that is, this portfolio had higher returns but at the cost of higher volatility. Its annual return was above 46%, but its diversified exposure to high-growth and high-volatility assets such as NVIDIA increased overall risk. This tradeoff is apparent in its positioning against the Nasdaq 100, with lower returns and, essentially, lower volatility.

In this respect, the strategy of annual rebalancing favored stability in the portfolio, as it corrected any drift in the weights regarding alignment with those from the optimized allocation. In any case, transaction costs were not included in this analysis; hence, it is left to future assessments that could include possible impacts.

The performance evaluation said that portfolio strategies should be in line with investor objectives and risk tolerance. High returns and Sharpe's ratio justify choosing a portfolio for long-term growth-oriented investors; volatility might be a relevant issue for the risk-averse investor who may opt for a more diversified strategy where the volatility is at a low level.

7 Conclusion

This research presents solid evidence of how AT can be utilized in portfolio management. It also discusses the challenges faced in optimizing risk-adjusted returns. The Sharpe ratio underlines the success of quantitative methods at 1.12, well above the Nasdaq 100 benchmark figure of 0.90. This corresponds to and articulates existing literature done by Sukma and Namahoot (2024), and Levendovszky et al. (2019), among other studies, which has shown the ability of algorithmic strategies to reduce heuristic biases and improve decision-making through data-driven methodologies.

The optimization-driven portfolio strategy favored high-performing assets; at 66.73%, NVIDIA comprised the biggest weighting, while it also dominated overall risk contribution at 77.09%. This demonstrates how the algorithm prioritizes assets with superior risk-return profiles, aligning with Mengshetti et al (2024) findings on multi-indicator performance strategies. This, however, concentrates much of the weight on one asset and underlines one of the main trade-offs between return maximization and diversification. According to Azhmyakov et al. (2023), closer attention should be paid to the increased susceptibility against specific market perils in case one wants to balance growth with stability in the portfolio.

With 40.89%, the portfolio volatility far exceeds that of the Nasdaq 100 at 21.47%, again proving that growth-oriented strategies alone face a host of challenges. While high volatility is usually characteristic of high-growth portfolios, sometimes it can unnerve the risk-averse investor (Courdent and McClelland 2022). These results indicate dynamic approaches that help mitigate risks without sacrificing return potential, especially in volatile markets. Apart from that, reliance on historical performance data carries the risk of overfitting; Lemma (2020) and Schmidt-Kessen et al. (2022) mentioned model fitting by historical market conditions that might generalize poorly in rapidly changing environments.

This therefore excludes some practical considerations like transaction costs, taxes, and liquidity constraints that are of most importance when assessing the realistic viability of an algorithmic strategy. According to Martins Pereira (2020) and Ramos and Perlin (2020), these factors are very important in assessing the real-world applicability of algorithmic strategies. For example, transaction costs could much lower profitability, especially in less liquid markets, whereas failure to consider after-tax consequences may distort net returns. Additional research is necessary to consider these aspects, providing a more comprehensive evaluation of algorithmic portfolio management.

Annual rebalancings were pretty efficient in keeping the weights near the target allocation, which agrees with what was expected by Yang et al. (2015) since they point out the importance of periodic adjustments to maintain the efficiency of a portfolio. On the other hand, frequent rebalancing may introduce extra transaction costs that offset the benefit arising from the reduced weight drift, as pointed out by Mukerji et al. (2019). Thus, the research into adaptive rebalancing methodologies—which would change with market conditions or volatility—may achieve an improved balance between cost efficiency and alignment of the portfolio.

In summary, this study contributes to the expanding literature that places AT at the center of the ongoing transformation in today's financial markets. For investors, the results suggest that algorithmic strategies are best suited for growth-oriented strategies that offer the sceptre of outperformance. Yet, such strategies also encapsulate certain risks, notably high concentration and volatility, which require active management and flexibility. These findings will provide industry practitioners with insight into refining trading strategies that incorporate practical constraints on transaction costs and liquidity conditions to enhance real-world applicability. Policymakers will learn from this research because it underlines the system-wide risks that could accrue due to algorithmic strategies, especially in the contexts of turbulent markets, in fact rendering the regulatory governance necessary so that a balance between market stability and innovation is achieved.

These results have more implications than portfolio management alone. By showing an information-based advantage in optimization systematically, our study puts algorithmic trading in the wider context of financial technology innovations and its increasing potential to cope with the rising complexity of the market environment. This replaces heuristics-based decision-making with objective methodologies—a theme that fits well within larger issues of automation and AI. However, Hassavayukul (2020) provides a warning that the effectiveness of these methods varies depending on market conditions and thus requires ongoing refinement and adjustment.

Limitations identified in this study, therefore, should be addressed in future studies by the inclusion of practical constraints, such as transaction costs, tax implications, and liquidity, in the evaluation of algorithmic strategies to make them more realistic. Dynamic rebalancing approaches that adjust frequency based on volatility or market conditions could further improve cost efficiency and risk management. This would potentially enrich the diversification and give a far better look into how algorithmic trading can operate in disparate markets from a wider range of asset classes: bonds, commodities, and alternative investments. The resulting portfolios would be exposed to very extreme market conditions or systemic shocks using stress testing, which could provide valuable insights into their resilience and performance under adverse scenarios.

References

Alaminos D, Salas MB, Callejón-Gil ÁM (2024) Managing extreme cryptocurrency volatility in algorithmic trading: EGARCH via genetic algorithms and neural networks. Quant Financ Econ 8(1):153–209

Azhmyakov V, Shirokov I, Dernov Y, Guzman Trujillo LA (2023) On a data-driven optimization approach to the pid-based algorithmic trading. J Risk Financ Manag 16(9):387. https://doi.org/10.3390/jrfm16090387

Cooper R, Ong M, Van Vliet B (2015) Multi-scale capability: a better approach to performance measurement for algorithmic trading. Algorithmic Finance 4(1–2):53–68. https://doi.org/10.3233/AF-150043

Courdent A, McClelland D (2022) The impact of algorithmic trading on market quality: evidence from the Johannesburg Stock Exchange. Invest Anal J 51(3):157–171. https://doi.org/10.1080/10293523.2022.2090056

Dananjayan MP, Gopakumar S, Narayanasamy P (2023) Unleashing the algorithmic frontier: navigating the impact of algo trading on investor portfolios. J Inf Technol Teach Cases 0(0)

Ersan O, Ekinci C (2016) Algorithmic and high-frequency trading in Borsa Istanbul. Borsa Istanbul Rev 16(4):233–248. https://doi.org/10.1016/j.bir.2016.09.005

Goyal H, Chandel N, Maurya A, Shailja (2024) AI-powered autonomous trading: enhancing market efficiency through predictive analytics. Int J Res Appl Sci Eng Technol 12(11):377–386. https://doi.org/10.22214/ijraset.2024.65001

Haddadian H, Baky Haskuee M, Zomorodian G (2022) A hybrid artificial intelligence approach to portfolio management. Iran J Financ 6(1):1–27. https://doi.org/10.30699/ijf.2021.287131.1237

Hassavayukul P (2020) Does algorithmic trading improve liquidity in emerging markets? Empirical evidence from Thailand. ABAC J 40(2):21–40

Jacob-Leal S, Hanaki N (2024) Algorithmic trading, what if it is just an illusion? Evidence from experimental asset markets. J Behav Exp Econ 112:102240. https://doi.org/10.1016/j.socec.2024.102240

Khurana SS, Singh P, Garg NK (2024) Og-cat: a novel algorithmic trading alternative to investment in crypto market. Comput Econ 63(5):1735–1756. https://doi.org/10.1007/s10614-023-10380-9

Kissell R (2011) TCA in the investment process: an overview. J Index Invest 2(1):60–64

Kraft D (1988) A software package for sequential quadratic programming. DFVLR-Forschungsbericht 88–28 German Aerospace Center (DLR) Institute for Flight Mechanics

Lemma V (2020) Innovations in asset management under a regulatory perspective: which rules can support the success of fintech? SSRN Electron J. https://doi.org/10.2139/ssrn.3657333

Levendovszky J, Reguly I, Olah A, Ceffer A (2019) Low complexity algorithmic trading by feed-forward neural networks. Comput Econ 54(1):267–279. https://doi.org/10.1007/s10614-017-9720-6

Martínez RG, García MLM, Martín FD (2022) Algorithmic trading based on the incidence of COVID-19 in Europe. Rev Econ Financ 20:536–545. https://doi.org/10.55365/1923.X2022.20.61

Martins Pereira C (2020) Unregulated algorithmic trading: testing the boundaries of the european union algorithmic trading regime. J Financ Regul 6(2):270–305. https://doi.org/10.1093/jfr/fjaa008

Mengshetti O, Gupta K, Zade N, Kotecha K, Mutha S, Joshi G (2024) Synergizing quantitative finance models and market microstructure analysis for enhanced algorithmic trading strategies. J Open Innov: Technol Mark Complex 10(3):100334. https://doi.org/10.1016/j.joitmc.2024.100334

Mestel R, Murg M, Theissen E (2018) Algorithmic trading and liquidity: long term evidence from Austria. Financ Res Lett 26:198–203. https://doi.org/10.1016/j.frl.2018.01.004

Mukerji P, Chung C, Walsh T, Xiong B (2019) The impact of algorithmic trading in a simulated asset market. J Risk Financ Manag 12(2):68. https://doi.org/10.3390/jrfm12020068

Muravyev D, Picard J (2022) Does trade clustering reduce trading costs? Evidence from periodicity in algorithmic trading. Financ Manage 51(4):1201–1229. https://doi.org/10.1111/fima.12405

Omran SM, El-Behaidy WH, Youssif AAA (2023) Optimization of cryptocurrency algorithmic trading strategies using the decomposition approach. Big Data Cogn Comput 7(4), Article 174. https://doi.org/10.3390/bdcc7040174

Patel SA (2023) Trading strategies using technical indicators: an active portfolio management approach. J Risk Financ Manag 26(1):138–151

Ramos HP, Perlin MS (2020) Does algorithmic trading harm liquidity? Evidence from Brazil. N Amn J Econ Financ 54, Article 101243. https://doi.org/10.1016/j.najef.2020.101243

Schmidt-Kessen MJ, Eenmaa H, Mitre M (2022) Machines that make and keep promises—lessons for contract automation from algorithmic trading on financial markets. Comput Law Secur Rev 46, Article 105717. https://doi.org/10.1016/j.clsr.2022.105717

Sharpe WF (1966) Mutual fund performance. J Bus 39(1):119–138

Sukma N, Namahoot CS (2024) Enhancing trading strategies: a multi-indicator analysis for profitable algorithmic trading. Comput Econ. https://doi.org/10.1007/s10614-024-10669-3

Sun S, He X, Wang R, Zhu J, Li J, An B (2021) DeepScalper: a risk-aware reinforcement learning framework to capture fleeting intraday trading opportunities. Digital Financ 5(1):1–2

Topiwala P, Dai W (2022) Surviving black swans: the challenge of market timing systems. J Risk Financ Manag 15(7):280

Weinan E, Ruimeng Hu, Peng Shige (2023) Deep Learning in Finance. Digital Financ 5:1–2. https://doi.org/10.1007/s42521-023-00080-2

Yang SY, Qiao Q, Beling PA, Scherer WT, Kirilenko AA (2015) Gaussian process-based algorithmic trading strategy identification. Quant Financ 15(10):1683–1703. https://doi.org/10.1080/14697688.2015.1011684

Zhou H, Kalev PS, Frino A (2020) Algorithmic trading in turbulent markets. Pac Basin Financ J 62:101358. https://doi.org/10.1016/j.pacfin.2020.101358

Assessing Bitcoin Return Extrema in the Context of Extreme Value Theory

Erhan Uluceviz

Abstract This study examines the distribution of extreme Bitcoin returns within the framework of Extreme Value Theory. Extreme Value Theory is crucial for modeling rare but impactful events in fields such as climate science, finance, insurance, engineering, and environmental sciences, among others. Using traded price data from Binance, the world's largest cryptocurrency exchange by volume, we analyze weekly extreme values (both maxima and negative minima) of hourly Bitcoin returns from January 2018 to July 2024. Specifically, we explore both tails of the Bitcoin hourly return distribution using the generalized extreme value class of distributions. Our results suggest that both tails can be effectively modeled with the generalized extreme value framework. This enables us to make hourly return predictions over weekly forecast horizons which can be instrumental in conducting risk assessments. Additionally, these insights could enhance the development of trading algorithms for this increasingly popular asset class, which is integrated already into traditional capital markets via exchange-traded funds (ETFs) on major stock exchanges.

Keywords Bitcoin returns · Extrema · Tail distributions

1 Introduction

This paper aims to demonstrate the application of Extreme Value Theory (EVT) in the cryptocurrency space, specifically for modeling the extremes of hourly Bitcoin returns, while also exploring its potential use in AI/ML applications. EVT has been instrumental in modeling rare and extreme events that can have significant consequences. It has two main components: (i) Block Maxima (or Minima) Method (BMM) and (ii) Peak Over Threshold (POT) Method. Generalized Extreme Value (GEV) distribution modeling is commonly used to model block maxima or minima. It was developed in the 1920s, with early contributions from Fisher and Tippett (1928), and

E. Uluceviz (✉)
Faculty of Business Administration, Gebze Technical University, Kocaeli, Türkiye
e-mail: erhan@uluceviz.com

M. Gün and B. Kartal (eds.), *Machine Learning in Finance*, Contributions to Finance and Accounting, https://doi.org/10.1007/978-3-031-83266-6_11

later rigorously formalized by Gnedenko (1943). The Generalized Pareto (GP) distribution is often used in POT modeling, focusing on modeling extremes that exceed a prespecified threshold. The foundational ideas that underlie it date back to papers by Pickands (1975) and Balkema and de Haan (1974).

Among the many applications of Extreme Value Theory across various fields, its use in quantitative risk management is widespread and is discussed in detail by McNeil et al. (2015). Several financial applications of EVT to Bitcoin price returns are also explored in the literature. Islam and Das (2021) apply EVT to predict long-run Bitcoin returns, using both the Block Maxima and POT methods. They find that the Block Maxima approach offers a better fit than the POT method. Osterrieder and Lorenz (2017) analyze the extreme value behavior of Bitcoin returns, emphasizing tail risk and comparing it to G10 currencies. Their findings show that Bitcoin exhibited volatility six to seven times higher than that of G10 currencies, along with significant tail risk. Ke et al. (2022) apply a dynamic POT model to forecast both lower and upper tail Value-at-Risk (VaR) for Bitcoin returns. Their findings show that the POT model outperforms GARCH-EVT models with various distributions, particularly in predicting lower tail risk, based on out-of-sample VaR forecasts. Zhang et al. (2019) use extreme value analysis to examine the tail risk behavior of hourly returns for Bitcoin, Ethereum, Litecoin, and Ripple. By estimating VaR and Expected Shortfall (ES) across varying thresholds, they find that Ripple is the riskiest cryptocurrency, exhibiting the largest potential gains and losses at all percentiles, while Bitcoin is the least risky. Koo et al. (2020) analyze the tail behavior of Bitcoin, Litecoin, Ethereum, and Ripple using the Autoregressive Fréchet model on five-minute high-frequency data. They find that Bitcoin exhibits significantly larger positive tail asymmetry, while the other three show a general tendency toward greater negative tail asymmetry in their intra-day returns. Additionally, all four cryptocurrencies exhibit lighter tails as the market matures.

In this paper, we apply the GEV distribution to model both tails of the Bitcoin hourly returns, specifically analyzing weekly block maxima and negative block minima. This dual approach allows us to examine both the upper and lower extremes of the return distribution, enabling us to identify and address any asymmetry in the behavior of extreme returns. Through GEV modeling, we find that the tail distributions of Bitcoin returns can be effectively modeled, opening up new research questions, particularly in areas such as risk management and asset trading. Successful density estimation paves the way for applications in risk management, including Value-at-Risk and Expected Shortfall. Our return level predictions, which estimate the expected weekly maximum and minimum over time horizons ranging from one quarter to 10 years, could also be utilized to develop trading algorithms. Additionally, our findings suggest that the dynamics of both tails are similar, providing insight into the underlying behavior of extreme Bitcoin returns.

The relationship between Extreme Value Theory and AI/ML is well-established. Although EVT is a methodology with roots dating back to the 1920s, it remains highly relevant for enhancing contemporary AI/ML applications, particularly in anomaly detection. Anomaly detection involves identifying patterns in data that deviate from expected or normal behavior, while extreme values, as modeled by

EVT, are located in the tails of the distribution (Chandola et al. 2009). Scheirer (2017) emphasizes the advantages of EVT for image recognition tasks, highlighting its strong statistical foundation and improved accuracy compared to Gaussian models, among other benefits. Additionally, extreme features are more important than average ones for effectively discriminating between objects. Rudd et al. (2018) introduced the Extreme Value Machine, an algorithm that leverages EVT to detect anomalies in training datasets with a minimal number of parameters. Vignotto and Engelke (2020) propose an anomaly detection algorithm based on EVT and offer a concise overview of the literature connecting EVT with AI/ML applications. While our work does not constitute a direct implementation of the above techniques, the high-frequency, highly volatile nature of our data, along with the modeling of tail behavior, includes elements that are relevant to specific domains within AI/ML contexts.

The rest of this paper is organized as follows. Section Two provides a brief literature review on EVT and its applications in various contexts. Section Three describes the dataset used in the analysis. Section Four presents the theoretical framework underlying the study. Section Five details the empirical results for both the block maxima and negative block minima series. Section Six discusses the findings, and finally, Section Seven concludes the paper.

2 Extreme Value Theory Literature in Various Contexts

While we discuss the major literature relevant to our analysis in the Introduction, this section focuses on EVT literature, particularly the block maxima approach, within different contexts.

Our review is by no means comprehensive. Excellent books and reviews are available such as Beirlant et al. (2004), Coles (2001), de Haan and Ferreira (2006), Embrechts et al. (1997), and Gumbel (1958) among others.

Methodological and statistical approaches in Extreme Value Theory (EVT) have been extensively studied. Some recent examples include Dombry (2015), who investigates the consistency and existence of maximum likelihood estimators in EVT using the block maxima method, and Dombry and Ferreira (2017), who discuss the properties of these estimators for block maxima in EVT. Faranda et al. (2011) examine the numerical convergence of the block maxima method to the generalized extreme value distribution. Ferreira and de Haan (2013) discuss block maxima methods and probability-weighted moment (PWM) estimators in EVT.

Bucher and Zhou (2018) compare two popular methods in extreme value theory: the block maxima method and the peak-over threshold approach. They find that the peak-over-threshold approach generally outperforms the block maxima method, providing more reliable estimates, especially when the tail distribution has a heavy tail.

The ability of extreme value theory to identify rare and extreme events has made it widely applicable in finance and trading. For instance, Liu et al. (2018) apply EVT to volatility forecasting, while Koliai (2016) uses EVT in stress testing to assess the resilience of trading strategies and the robustness of risk management systems. Additionally, Muelaa et al. (2017) show that, in terms of accuracy for Value-at-Risk estimations and daily capital requirements, conditional extreme value theory outperforms the parametric method. Fretheim and Kristiansen (2015) apply EVT to food futures prices to assess whether extreme price volatility has increased over time and, as a result, find no evidence of a rise in extreme risk.

There is a wealth of literature that applies EVT and GEV modeling to various applications. For instance, Ali et al. (2023) use the block maxima approach to estimate the tail behavior of safety-related metrics during lane-changing events. Osei et al. (2021) apply the Gumbel distribution to estimate the return periods of rainfall and floods in the Pra River Catchment in Ghana, providing critical estimates for effective flood management and infrastructure planning. Ragno et al. (2019) introduce a novel framework for analyzing nonstationary extreme values in hydrological processes, such as extreme rainfall, floods, or droughts. The authors develop a nonstationary generalization of the extreme value distributions that can adapt to changing conditions. Soukissian and Tsalis (2015) investigate how different methods of estimating parameters for the generalized extreme value (GEV) distribution affect the prediction of extreme wind speeds in real-world settings.

3 Dataset

Our dataset is derived from raw BTC/USDT prices provided by Binance.com.[1] Using its application programming interface (API), Binance offers open, high, low, and close price data, along with trading volume, at a 1-min frequency for all crypto assets traded on its platform since 2017-08-17 04:00:00.[2] To address potential microstructure effects, we aggregated the 1-min closing data to a 1-h frequency and calculated log returns based on the hourly closing prices. The typical data frequency recommended in market microstructure literature is 5 min (Hansen and Lunde, 2006; Ait-Sahalia and Jacod, 2014). However, because the Bitcoin market is less efficient than more developed markets such as foreign exchange or stock markets, we opted to use a 1-h frequency. Given that Binance launched in Summer 2017 and experienced operational challenges during 2017, the available data contains some gaps, and the prices are assumed to be less representative of global prices due to low liquidity; therefore, we decided to exclude data before 2018-01-01 and we focus on the hourly data from 2018-01-01 00:00:00 to 2024-07-31 23:00:00. Our timestamps correspond to the beginning of the hour that they represent.

[1] Binance is the world's largest cryptoexchange. As of 2024-10-29, 24 h volume on Binance is about USD 17.7 billion. Source: https://coinmarketcap.com/rankings/exchanges/, accessed on 2024-10-29.

[2] For consistency, timestamps in this paper are formatted as YYYY-MM-DD HH:MM:SS and are provided in UTC.

Based on the raw data we have, we created weekly block maxima and negative block minima series. We will explain the use of the block maxima and negative block minima series when we discuss EVT methodology in detail in Sect. 4.

Table 1 presents descriptive statistics for the series which form the basis of our analysis. We report 11 different statistics for each time series. The first data column displays our raw data series, consisting of 57,549 hourly return observations. This data exhibits a negative skewness of −0.5157 and a kurtosis of 37.8059. The Jarque–Bera statistic exceeds 3.4 million and is statistically significant at the 1% level. These findings suggest that our raw data series is not normally distributed and has fat tails. This suggests that using non-normal approaches like extreme value theory to model the tails or extremes is empirically justified. Figure 1 presents a QQ plot that compares the theoretical quantiles of a normal distribution (with its mean and variance estimated from the 1-h return series) against the sample quantiles of the same series. Both tails of the series deviate from the 45-degree line substantially, providing visual evidence of the series' non-normality.

To model the weekly maxima of the raw series as input for our modeling, detailed in Sect. 4, we construct a 1-week block maxima series. This series is formed by selecting the maximum observations for each week. From the available hourly observations, we derive 344 weekly maximum values. These observations have a skewness of 2.5017 and a kurtosis of 10.0376, with a Jarque–Bera statistic of 1,803, which is significant at the 1% level. Together, these statistics indicate that the weekly maxima series is also non-normally distributed.

Similarly, to model the weekly minima of the series, we create a 1-week negative block minima series that captures the minimum values for each week. Since EVT primarily focuses on modeling maxima, the standard approach for modeling minima involves analyzing the maxima of the negative block minima. This approach enables

Table 1 Descriptive statistics for Bitcoin returns

Statistic	1-Hour log returns	1-Week block maxima log returns	1-Week negative block minima log returns
Count	57,549	344	344
Mean	0.0000	0.0285	0.0296
Std	0.0077	0.0187	0.0194
Min	− 0.2010	0.0049	0.0060
1Q	− 0.0025	0.0164	0.0179
Median	0.0001	0.0241	0.0244
3Q	0.0027	0.0342	0.0364
Max	0.1603	0.1603	0.2010
Skewness	− 0.5157	2.5017	2.9558
Kurtosis	37.8059	10.0376	17.8760
Jarque–Bera statistics	3,429,807*	1,803*	5,081*

*Statistical significance at the 1% level

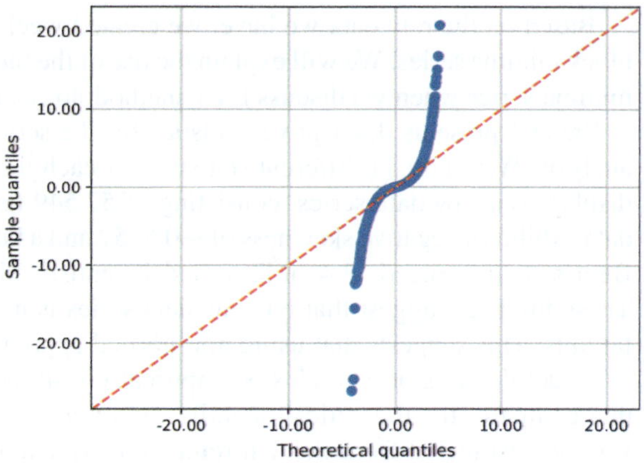

Fig. 1 QQ plot for 1-hour return series

us to apply maxima modeling techniques to minima. The descriptive statistics for the negative block maxima series also indicate non-normal behavior, with a skewness of 2.9558, a kurtosis of 17.8760, and a Jarque–Bera statistic of 5,081, which is significant at the 1% level.

With these observations and arguments in mind, we are now ready to model the extremes of Bitcoin returns using GEV distributions within the EVT framework, as discussed in Sect. 4.

4 Generalized Extreme Value Distribution Within the EVT Framework

In the preface of their book; Embrechts etal. (1997) highlight that extreme value theory effectively leverages data on extreme phenomena, offering a more robust approach than empirical curve fitting or guesswork. They caution that, in the absence of reliable methods like EVT, practitioners may resort to less credible alternatives. In this respect, the generalized extreme value distribution, as part of extreme value theory, is specifically designed to model rare, low-probability events that can lead to catastrophic consequences.

This section provides a concise overview of the relevant aspects of extreme value theory, focusing specifically on the generalized extreme value distribution. For a more comprehensive discussion, see e.g. Coles (2001) or Embrechts etal. (1997).

Consider a sequence of n independent and identically distributed (i.i.d.) random variables $X_1, X_2, \ldots, X_n$ with a common distribution function F. In this context, X_i could represent an hourly return series. Let M_n be the maximum of these n observations, which could correspond to a week's worth of data. The independence of the X_i values leads to the distribution of M_n:

$$P(M_n \leq s) = P(X_1 \leq s).P(X_2 \leq s)\ldots P(X_n \leq s) = F_n(s) = [F(s)]^n \quad (1)$$

As n → ∞, limiting distribution of $F_n(s) \to 0$, if $F(s) < 1$ or $F_n(s) \to 1$, if $F(s) = 1$. To ensure that F_n is not degenerate, one can apply a normalization of M_n (that is M_n^*) to guarantee convergence to a non-degenerate distribution G(s):

$$M_n^* = \frac{M_n - b_n}{a_n}, \tag{2}$$

For sequences of constants $a_n > 0$ and b_n.

Fisher–Tippett–Gnedenko theorem[3] states that if sequences of constants $a_n > 0$ and b_n exist, as $n \to \infty$, such that

$$P\left(M_n^* = \frac{M_n - b_n}{a_n} \le s\right) \to G(s) \tag{3}$$

where $G(s)$ is a non-degenerate distribution function, then the limiting distribution $G(s)$ will belong to one of the following families: (i) Gumbel, (ii) Fréchet, or (iii) Weibull.

The unification of these three types of distributions into a single family, as proposed by Jenkinson (1955), leads to the GEV distribution, a three-parameter distribution of the form:

$$G(s) = exp\left\{-\left[1 + \xi\left(\frac{s - \mu}{\sigma}\right)\right]^{-\frac{1}{\xi}}\right\} \tag{4}$$

which is defined for $1 + \xi\left(\frac{s-\mu}{\sigma}\right) > 0$. Here, μ represents the location parameter, σ is the scale parameter and ξ is the shape parameter. The parameters satisfy $\mu \in \mathbb{R}$, $\sigma > 0$ and $\xi \in \mathbb{R}$. When $\xi \to 0$, the GEV distribution simplifies to the Gumbel distribution. The Fréchet distribution is obtained when $\xi > 0$ while the Weibull distribution arises when $\xi < 0$.

It's important to note that, similar to the Central Limit Theorem (CLT), which justifies using the normal distribution to approximate averages, Extreme Value Theory (EVT) provides a solid theoretical framework that stabilizes the distributions of various random variables. This stabilization allows their maxima to asymptotically converge to specific limiting distributions. In this way, EVT serves as the extreme value analog of the CLT (Coles 2001).

[3] While the limiting laws for maxima were derived by Fisher and Tippett (1928), Gnedenko (1943) provides a first rigorous proof.

5 Empirical Approach and the Results

This section presents the empirical results of our analysis. The first subsection presents the results of the block maxima analysis, while the second subsection provides the results for the negative block minima in the GEV estimation of Bitcoin returns.

To estimate the model parameters, we utilized scipy.stats.genextreme, a class in the Python SciPy library's statistics module. This class represents the Generalized Extreme Value distribution and provides methods for the necessary computations.

To compute the parameter estimates for our model, we employed two approaches: first, the standard Maximum Likelihood Estimation (MLE) using the scipy.stats.genextreme class in Python, and second, a bootstrapping method developed by Efron (1979). Although our raw dataset is a time series object, as discussed in Sect. 4, the GEV model relies on the assumption of i.i.d. observations. Hourly maxima (or negative minima) of weekly blocks can also be considered to follow the i.i.d. assumption. Therefore, instead of using time-series bootstrapping, which involves moving blocks of consecutive observations (Künsch 1989), we created bootstrap samples using the most straightforward method: simple random sampling with replacement. We conducted 2,000 bootstrap replications for both the block maxima and negative block minima analyses. In the following, we present the results from both approaches. In all of our results, the significance level is set at 5%.

5.1 Block Maxima Series: Estimation Results

The estimation results for the block maxima approach are presented in Table 2. To maintain consistency and save space, we only provide the bootstrap-based results. The MLE-based coefficient estimates and model fit show minimal differences.[4]

From Table 2, with a negative shape parameter estimate $\hat{\xi} = -0.2438 < 0$, we conclude that the estimated model follows a Weibull distribution. To further assess the fit of our model, we conducted a Kolmogorov–Smirnov (KS) test to evaluate whether the data follows the GEV distribution. The KS statistic is computed to be 0.0436, with a p-value of 0.8999. Therefore, we fail to reject the null hypothesis, indicating that the data is consistent with the GEV distribution.

In Table 2, we also present return levels (z_p) associated with return periods[5] $\left(\frac{1}{p}\right)$ which represent the levels expected to be exceeded each week with a probability p. In EVT parlance, the weekly maximum is expected to exceed z_p in any given week with a probability of p, i.e. over every p period, the maximum is expected to exceed z_p. We provide return levels for 13, 26, 52, 104, and 520 weeks, along with their

[4] The results are available upon request from the author.

[5] The return period refers to the expected time between extreme events, as discussed by Embrechts et al. (1997). While we apply this concept in a financial context, it is a generic term that can be used across various fields.

Table 2 Estimation results for block maxima series

Parameters	Estimate (s.e.)	Confidence interval
Shape ($\hat{\xi}$)	– 0.2438 (0.1174)	(– 0.7741, – 0.1439)
Location ($\hat{\mu}$)	0.0198 (0.0006)	(0.0186, 0.0212)
Scale ($\hat{\sigma}$)	0.0103 (0.0009)	(0.0092, 0.0130)
Return periods	*Level*	*Confidence interval*
13-W	0.0574	(0.0494, 0.1262)
26-W	0.0749	(0.0603, 0.2178)
52-W	0.0978	(0.0723, 0.3741)
104-W	0.1296	(0.0855, 0.6410)
520-W	0.2785	(0.1218, 2.2283)

corresponding confidence intervals. Since we are working with hourly data, the return levels associated with weekly maxima refer to hourly values. Investors can therefore expect the maximum hourly values within the specified confidence intervals over various time horizons, ranging from one quarter (13 weeks) to 10 years (520 weeks). Confidence intervals are determined using the 2.5th and 97.5th percentiles of the estimated return level values. Our return level results suggest that, for example, over a 13-week horizon, one could expect a block maximum return of 5.74%, with a 95% confidence interval ranging from 4.94% to 12.62%. One potential use of return levels for block maxima is in asset trading. These values can serve as take-profit levels for assets under management. However, it is important to note that significant backtesting and model validation would be required to effectively incorporate this approach into a trading strategy.

As a final step, we present Fig. 2, which consists of two panels. In panel (a), we display a QQ plot, where the theoretical quantiles correspond to our fitted GEV model. In panel (b), we show the fitted probability density function (PDF) of the GEV model alongside the histogram of the raw data.

In panel (a), we observe that nearly all of the observations align closely with the 45-degree line, indicating a strong fit of the model to the data. Only a very few extreme observations show slight misalignment. Similarly, in panel (b) of Fig. 2, the fitted GEV model closely matches the data. Together, the two panels of Fig. 2 validate the effectiveness of our approach in modeling the weekly maxima of Bitcoin returns.

5.2 Negative Block Minima Series: Estimation Results

The estimation results for the negative block minima approach are presented in Table 3. The rationale for presenting the results and the other details of the analysis follows the same approach as in Sect. 5.1. Therefore, in this subsection, we focus on

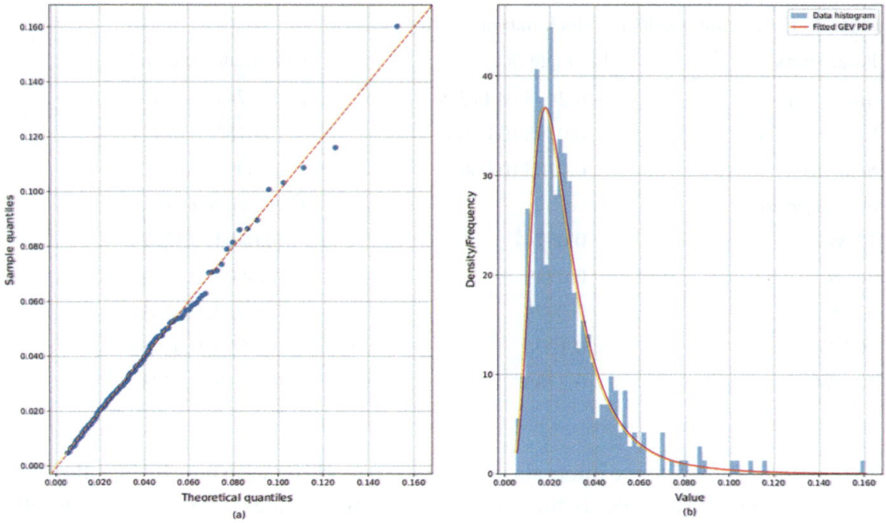

Fig. 2 Weekly block maxima fit a GEV model

discussing the results without reiterating the modeling and presentation choices. As before, we present only the bootstrap-based results.

With a negative shape parameter estimate $\hat{\xi} = -0.2572 < 0$, the estimated model is of a Weibull type. The KS statistic is 0.0407, with a p-value of 0.9387, and hence indicates that the data is consistent with the GEV distribution.

When comparing the estimated model parameters for both models in Tables 2 and 3, we observe that all three parameters are relatively similar. This suggests that the behavior of both tails may be comparable. However, further analysis is required to draw a definitive conclusion.

In terms of return levels and return periods, our results suggest that, for example, over a 13-week horizon, one could expect an hourly return of – 5.91%, with a 95%

Table 3 Estimation results for negative block minima series

Parameters	Estimate (s.e.)	Confidence interval
Shape ($\hat{\xi}$)	– 0.2572 (0.1093)	(– 0.6546, – 0.1475)
Location ($\hat{\mu}$)	0.0206 (0.0007)	(0.0192, 0.0219)
Scale ($\hat{\sigma}$)	0.0105 (0.0008)	(0.0094, 0.0123)
Return periods	*Level*	*Confidence interval*
13-W	0.0591	(0.0513, 0.0999)
26-W	0.0769	(0.0629, 0.1579)
52-W	0.0999	(0.0758, 0.2489)
104-W	0.1308	(0.0901, 0.3923)
520-W	0.2666	(0.1287, 1.1253)

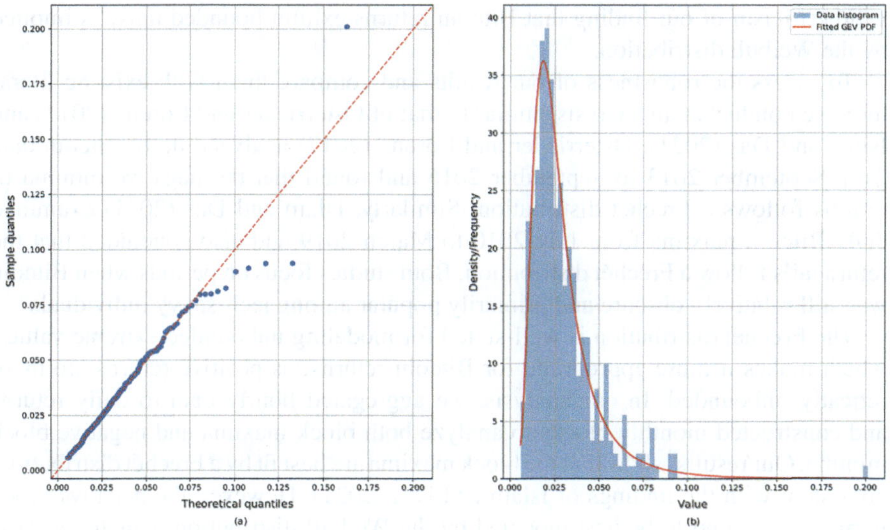

Fig. 3 Weekly negative block minima fit a GEV model

confidence interval ranging from − 9.99 to − 5.13%. Since we are analyzing the maxima of the negative block minima series, the interpretation should focus on the negative values.

An important potential use of return levels for negative block minima is in risk and investment management, or for setting stop-loss levels in asset trading. These return levels can help identify extreme negative events, which are crucial for managing potential losses and making informed trading decisions. These efforts fall outside the scope of the current project and will therefore be explored as extensions in future work.

To visually evaluate the performance of our fit, we present Fig. 3. In the QQ plot of panel (a), we observe that, except for a few observations (particularly those in the 7.5–10.0% range in the sample quantiles), most of the data points align closely with the 45-degree line. This indicates a strong fit of the model to the data.

Similarly, as shown in panel (b) of Fig. 3, the PDF of the GEV model closely matches the data.

6 Discussion

The GEV analysis of Bitcoin returns suggests robust tail behavior, with the negative shape parameters indicating a Weibull distribution (as discussed in Sects. 5.1 and 5.2). The Weibull distribution, which is bounded, is commonly used to model non-negative variables such as lifetimes or times of failure. However, Bitcoin returns, especially on an hourly basis, are typically considered unbounded. This raises a valid criticism of our finding that Bitcoin returns exhibit bounded tails, as implied by the Weibull distribution.

To assess the robustness of our results and compare them with existing literature, we conducted an analysis similar to that of Osterrieder and Lorenz (2017) and Islam and Das (2021). Osterrieder and Lorenz (2017) analyzed daily Bitcoin data from September 2013 to September 2016 and found that the negative minima of returns follows a Frechét distribution. Similarly, Islam and Das (2021) examined daily Bitcoin maxima from July 2010 to March 2019 and also concluded that the return tails follow a Frechét distribution. Both studies focus on periods when Bitcoin was still relatively obscure and primarily popular among tech-savvy individuals.

The Frechét distribution is well-suited for modeling unbounded extreme values, which makes it more appropriate for Bitcoin returns, as positive returns are theoretically unbounded. In our analysis, we aggregated hourly data to daily returns and constructed monthly blocks to analyze both block maxima and negative block minima. Our results[6] show that the block maxima are best fit by a Frechét distribution, consistent with the findings of Islam and Das (2021). However, the negative block minima are found to be best modeled by the Weibull distribution, which contrasts with the results of Osterrieder and Lorenz (2017).

Notably, there are differences in the data and data ranges between our study and those of Islam and Das (2021) and Osterrieder and Lorenz (2017). Islam and Das (2021) report a data range from − 57 to 336%, while our daily data range from − 50 to 18%. The -50% return in our dataset occurred on 2020-03-12, amid market turbulence driven by fears of the impending COVID-19 pandemic and a widespread global sell-off across all asset classes. Neither of the previous studies covers this period. This suggests that, relative to the data used by Islam and Das (2021), our dataset has a smaller dispersion and could be considered more bounded, which may help explain the differences in the distributional fits.

In addition, our data are sourced exclusively from Binance.com, which provides its trade data, whereas the data used by the other studies were aggregated from multiple platforms. We chose Binance.com primarily because it offers publicly available, consistent, free datasets, unlike other vendors that charge substantial fees for high-frequency data.

Bitcoin has also evolved into a more prominent asset class. It is now represented by Exchange-Traded Funds (ETFs) traded on major stock exchanges, and it has become part of the portfolios of hedge funds and individual investors seeking the security of regulated exchanges, with oversight from entities like the U.S. Securities and Exchange Commission (SEC).[7] The growing maturity of Bitcoin as an investment asset may also contribute to differences in our findings compared to earlier studies.

---

[6] To save space, we do not report the full results here; however, they are available upon request from the author.

[7] On 2024-01-10, SEC chairman Gary Gensler stated, "The Commission approved the listing and trading of a number of spot Bitcoin exchange-traded product (ETP) shares." Source: https://www.sec.gov/newsroom/speeches-statements/gensler-statement-spot-bitcoin-011023, accessed on 2024-11-16.

In conclusion, our findings are largely consistent with the existing literature but provide a more recent analysis using high-frequency data. The differences in date range, data source, and the evolving role of Bitcoin as an investment asset highlight the importance of considering these factors when modeling its return distributions.

7 Conclusion

This study aims to model the weekly extremes (both maxima and negative minima) of hourly Bitcoin returns from January 2018 to July 2024 within the framework of Extreme Value Theory. We utilize traded price data from Binance, the world's largest cryptocurrency exchange by traded volume.

Within the framework of Extreme Value Theory, we used the Generalized Extreme Value distribution as our primary modeling tool, which is estimated through bootstrapping. The Generalized Extreme Value distribution is widely employed to model extreme events across various fields, including climate science, finance, insurance, engineering, environmental sciences, and hydrology, among others.

The application of EVT to cryptocurrency markets is relatively limited, and the existing studies typically rely on daily data, often focusing on modeling only one tail of the distribution. In this regard, our use of high-frequency (hourly) data and the modeling of both tails represents a significant contribution to the field.

We find that our hourly raw data exhibit strong non-normal behavior, consistent with findings widely documented in the literature. By focusing on the maxima (and negative minima) of the series as the basis for analysis, we can successfully model the tail distributions. Our findings suggest that the model parameters for both the maxima and minima are similar, providing evidence for the comparable behavior of both tail distributions. This observation represents a novel finding, to the best of our knowledge.

The main limitation of this research is the lack of cross-validation with other well-established methodologies, such as GARCH models. However, our findings pave the way for further research and development in the field. Specifically, our approach creates new opportunities, particularly in cryptocurrency risk management, investment strategies, and trading.

As prospective extensions of this project, we envision applications of Value-at-Risk (VaR) or Expected Shortfall (ES) in the risk management domain. Weekly, daily, or hourly block analyses could also be used in backtests in trading applications.

References

Ali Y, Haque M, Zheng Z (2023) Assessing a connected environment's safety impact during mandatory lane-changing: a block maxima approach. IEEE Trans Intell Transp Syst 24:6639–6649

Ait-Sahalia Y, Jacod J (2014) High-frequency financial econometrics. Princeton University Press

Beirlant J, Goegebeur Y, Segers J, Teugels J (2004) Statistics of extremes: theory and applications. Wiley

Balkema AA, de Haan L (1974) Residual life time at great age. Ann Probab 2(5):792–804

Bucher A, Zhou C (2018) A horse race between the block maxima method and the peak–over–threshold approach. Stat Sci 36(3):360–378

Chandola V, Banerjee A, Kumar V (2009) Anomaly detection: a survey. ACM Comput Surv 41(3):1–58

Coles S (2001) An introduction to statistical modeling of extreme values. Springer

de Haan L, Ferreira A (2006) Extreme value theory: an introduction. Springer

Dombry C (2015) Existence and consistency of the maximum likelihood estimators for the extreme value index within the block maxima framework. Bernoulli 21(1):420–436

Dombry C, Ferreira A (2017) Maximum likelihood estimators based on the block maxima method. Bernoulli 23(3):1956–1984

Efron B (1979) Bootstrap methods: another look at the jackknife. Ann Stat 7(1):1–26

Embrechts P, Klüppelberg C, Mikosch T (1997) Modelling extremal events for insurance and finance. Springer

Faranda D, Lucarini V, Turchetti G, Vaienti S (2011) Numerical convergence of the block-maxima approach to the generalized extreme value distribution. J Stat Phys 145(6):1156–1180

Ferreira A, de Haan L (2013) On the block maxima method in extreme value theory: PWM estimators. Ann Stat 43:276–298

Fisher RA, Tippett LHC (1928) Limiting forms of the frequency distribution of the largest or smallest member of a sample. Math Proc Cambridge Philos Soc 24(2):180–190

Fretheim T, Kristiansen G (2015) Commodity market risk from 1995 to 2013: an extreme value theory approach. Appl Econ 47:2768–2782

Gnedenko BV (1943) Sur la distribution limite du plus grand de n valeurs [On the limiting distribution of the maximum term in a random series]. Ann Math 44(3):423–453

Gumbel EJ (1958) Statistics of extremes. Columbia University Press

Hansen PR, Lunde A (2006) Realized variance and market microstructure noise. J Bus Econ Stat 24(2):127–161

Islam MT, Das KP (2021) Predicting Bitcoin return using extreme value theory. Am J Math Manag Sci 44(4):315–329

Jenkinson AF (1955) The frequency distribution of the annual maximum (or minimum) values of meteorological elements. Q J R Meteorol Soc 81:158–171

Koliai L (2016) Extreme risk modeling: an EVT–pair-copulas approach for financial stress tests. J Bank Finance 70:1–22

Ke R, Yang L, Tan C (2022) Forecasting tail risk for Bitcoin: a dynamic peak over threshold approach. Financ Res Lett 49

Koo CK, Semeyutin A, Lau MCK, Fu J (2020) An application of autoregressive extreme value theory to cryptocurrencies. Singap Econ Rev 65(2):1–8

Künsch HR (1989) The jackknife and the bootstrap for general stationary observations. Ann Stat 17(3):1217–1241

Liu G, Wei Y, Chen Y, Yu J, Hu Y (2018) Forecasting the value-at-risk of the Chinese stock market using the HARQ model and extreme value theory. Physica A 505:395–404

McNeil AJ, Frey R, Embrechts P (2015) Quantitative risk management: concepts, techniques, and tools. Princeton University Press

Muelaa SB, Martínb CL, Sanzc RA (2017) An application of extreme value theory in estimating liquidity risk. Eur Res Manag Bus Econ 23:157–164

Osei M, Amekudzi L, Omari-Sasu AY, Yamba EI, Quansah E, Aryee J, Preko K (2021) Estimation of the return periods of maxima rainfall and floods at the Pra River Catchment, Ghana, West Africa using the Gumbel extreme value theory. Heliyon 7(5)

Osterrieder J, Lorenz J (2017) A statistical risk assessment of Bitcoin and its extreme tail behavior. Ann Financ Econ 12(1)

Pickands J (1975) Statistical inference using extreme order statistics. Ann Stat 3(1):119–131

Ragno E, Aghakouchak A, Cheng L, Sadegh M (2019) A generalized framework for process-informed nonstationary extreme value analysis. Adv Water Resour 128:51–66
Rudd EM, Jain LP, Scheirer WJ, Boult TE (2018) The extreme value machine. IEEE Trans Pattern Anal Mach Intell 40(3):739–746
Soukissian T, Tsalis C (2015) The effect of the generalized extreme value distribution parameter estimation methods in extreme wind speed prediction. Nat Hazards 78:1777–1809
Scheirer WJ (2017) Extreme value theory-based methods for visual recognition. Synth Lect Comput Vis 7(1)
Vignotto E, Engelke S (2020) Extreme value theory for anomaly detection—the GPD classifier. Extremes 23(4):501–520
Zhang Y, Chan S, Nadarajah S (2019) Extreme value analysis of high-frequency cryptocurrencies. High Frequency 2(1):61–69

Machine Learning in Portfolio Optimization

Diler Türkoğlu and Melih Kutlu

Abstract This study aims to compare the traditional mean–variance model with Firefly and Simulated Annealing algorithm in terms of portfolio performance. For this purpose, price data of BIST Dividend 25 Index stocks between 2018–2023 were used. Expected return, portfolio risk, Sharpe Ratio, Coefficient of variation, and Downside Risk were used for performance. The Firefly and Simulated Annealing algorithm portfolios generally have higher return rates than average and therefore carry more risk. The Firefly portfolio generally performs well in Sharp Ratio and Downside Risk. Investors can diversify with Firefly and SAs and achieve excess-market returns in Borsa İstanbul.

Keywords Portfolio optimization · Metaheuristic algorithm · Firefly algorithm · Simulated annealing

1 Introduction

Portfolio optimization continues to develop with the Mean-Variance (MV) approach. It is a subject that forms the backbone of investment theory after the work of Markowitz (1952). Different views have been put forward to overcome the limitations of the MV approach. The development of the investment environment with globalization has increased the search for more useful and more efficient portfolio optimization applications.

The fundamental premise underlying portfolio management is the notion that investors seek to maximize their returns while minimizing the associated risks (Butler

D. Türkoğlu (✉)
The Faculty of Economics and Administrative Sciences, Business Administration, Ardahan University, Ardahan, Türkiye
e-mail: diler.turkoglu@ardahan.edu.tr

M. Kutlu
The Faculty of Political Sciences, International Trade and Business, Samsun University, Samsun, Türkiye
e-mail: melih.kutlu@samsun.edu.tr

M. Gün and B. Kartal (eds.), *Machine Learning in Finance*, Contributions to Finance and Accounting, https://doi.org/10.1007/978-3-031-83266-6_12

and Lambson 2018). This principle is encapsulated in the Markowitz formulation of portfolio analysis, which posits that investors' preferences are determined by the mean and variance of their portfolio's returns. (Jobson and Korkie 1980) The Markowitz theory revolutionized the way investors think about constructing portfolios by recognizing that an asset's risk and return should be evaluated in the context of the entire portfolio, rather than in isolation (Draviam and Chellathurai 2002).

Portfolio management is a dynamic process. With developing technologies, new perspectives are being gained in portfolio management. The financial industry has been significantly impacted by artificial intelligence (AI) in recent years. The negativities experienced while creating a portfolio can be eliminated with artificial intelligence algorithms (Sutiene et al. 2024).

The performance of these two algorithms in optimizing portfolio selection was investigated comparatively with the traditional Markowitz MV model in this chapter. Portfolio optimization is a critical task in finance, where the goal is to identify the combination of financial assets that best meets an investor's objectives, often by balancing risk and return. We explore the use of two powerful optimization algorithms, the Firefly Algorithm (FA) and the Simulated Annealing algorithm (SA), in the context of portfolio optimization. Portfolio optimization is an important tool in investment. This optimization aims to keep risk and return at an optimum level. Investors have goals and portfolios should be created in line with these goals. We will try to achieve this with two algorithms. FA, a metaheuristic algorithm inspired by nature that mimics the blinking behavior of fireflies (Johari et al. 2013), has been successfully applied to a wide range of optimization problems, including those in finance (Tuba and Bačanin 2014). The FA is successful in multiple global optimization problems (Yang and He 2013). SA is a general probabilistic algorithm used for global optimization problems (Monticelli et al. 2007; Xiang et al. 2013).

We conducted our research on the Borsa Istanbul (BIST) Dividend 25 index. BIST Dividend 25 Index includes 25 companies with high dividend yields and high market value of the publicly traded parts of the company. These stocks are liquid. BIST Dividend 25 Index has a structure that guides stock investors who care about both dividends and liquidity (Mazgit 2013). On the other hand, the index includes companies from very different sectors. This will contribute to the diversification that will be done through algorithms.

The rest of this chapter is organized as follows: Section Two provides a general framework of portfolio management; Section Three presents the literature about the topic; Section Four outlines the methodology, and data structure, and provides findings; and Section Five involves a discussion of the findings and the conclusion.

2 General Framework and Criticism of Markowitz Portfolio Optimization

Portfolio management has long been a critical aspect of investment strategy, as investors seek to optimize their returns while managing risk (Constantinides and Malliaris 1995). The fundamental challenge for equity investors is to construct a stock portfolio that maximizes expected returns subject to acceptable risk levels (Lyle and Yohn 2021; Chaves et al. 2011; Stoyanov et al. 2007).

Markowitz's pioneering work in the 1950s introduced the mean–variance framework that formalized the risk-return trade-off in portfolio selection (Fabozzi et al. 2012; Avramov and Zhou 2010; Keller and Putten 2013). This model aims to minimize the investor's expected return and the level of risk he targets (Kaplan and Siegel 1994; Bailey et al. 1998). This is achieved through diversification, and portfolio variance is reduced by the assets included in the portfolio.

Diversification is the key to effective portfolio management, as it enables investors to mitigate the impact of individual asset risks by spreading their investments across a range of uncorrelated or negatively correlated assets. According to Markowitz (1999), Shakespeare mentioned diversification many years ago without knowing its importance:

My ventures are not in one bottom trusted,
Nor to one place; nor is my whole estate.
Upon the fortune of this present year;
Therefore, my merchandise makes me not sad.
(Merchant of Venice; Year: 1598; Shakespeare W.; Act I, Scene 1).

The mean–variance approach of Markowitz has become widely accepted by both academics and practitioners (Jensen 1972; Markowitz 1991). It represents a strategic, long-term approach to asset allocation, in contrast to more tactical, short-term momentum-based strategies (Markowitz and Dijk 2003; Jensen 1972). The key insight is that by diversifying investments, investors can reduce portfolio risk without necessarily sacrificing expected returns (Keller and Putten 2013). While the Markowitz framework provides a powerful conceptual foundation for portfolio management, its practical application faces several significant challenges (Rubinstein 2002).

Empirical studies have provided valuable insights into the performance of portfolio management strategies (Kolm et al. 2013). Markowitz's pioneering work has been extended to incorporate additional real-world constraints, such as cardinality limits on the number of assets and quantity limits on the investment in each asset (Baumöl 1966).

The trust problem experienced in historical data and the inconsistency of estimates based on this data cause the desired performance level to be achieved in portfolios (Kan and Zhou 2007; Hurley and Brimberg 2015; Behr et al. 2013). There are restrictions on the number of assets in the portfolio (cardinality restrictions) or the amount invested in each asset (amount restrictions) (Cai et al. 2020). We cannot reveal the real market structure with these restrictions (Aït-Sahalia et al. 2015). Researchers are constantly trying to improve the model to eliminate these deficiencies (Xiong and Idzorek 2011).

Markowitz portfolio optimization may present problems in estimating the basic parameters due to its dimensional and computational complexity (Avramov and Zhou 2010; Zhu et al. 2011). The Markowitz model has been used in many studies, but there is also a great demand for its development (Fernández-Navarro et al. 2021). The mean–variance approach assumes that asset returns follow a normal distribution. It does not account for the skewness and fat tails commonly observed in return distributions (Xiong and Idzorek 2011; Nisani 2018). The Markowitz model uses a variance–covariance matrix. However, this matrix also has certain difficulties in estimation (Hurley and Brimberg 2015; Zhu et al. 2011).

Characteristics of real-world portfolio problems, such as their large scale, strict real-world requirements, limited computation time, and imprecise parameter estimates, can make analytical optimization methods less suitable (Lyle and Yohn 2021). As a result, researchers and practitioners have turned to heuristic techniques, such as metaheuristic algorithms, to find high-quality solutions in a reasonable timeframe. (Abdel-Basset et al. 2018; Boussaïd et al. 2013). These advances in algorithms related to portfolio optimization have helped bridge the gap between theory and practice.

3 Literature Review

There are efforts in the literature to develop the FA with different algorithms. Tuba and Bacanin (2014) compared the modified FA, particle swarm optimization (PSO), and genetic algorithm (GA) over portfolios. The study was conducted on the weekly returns of stocks in the Hong Kong Hang Seng, German DAX 100, British FTSE 100, US S&P 100, and Japanese Nikkei indices. The modified FA performed better in terms of diversification. This algorithm offers a more balanced investment strategy. In another study that combined the FA with the artificial bee colony algorithm, Tuba and Bacanin (2014) found that the combined model performed much better than the classical artificial bee colony in the study he conducted with Tuba and Bacanin (2014) on the same indices.

Heidari and Neshatizadeh (2018), compared the FA and the Imperialist Competitive Algorithm (ICA) using daily stock price data of 25 companies listed on the Tehran Stock Exchange in the period 2010–2016 and stated that both algorithms performed well. In another study conducted for the Indonesian Stock Exchange, Lazulfa (2019) tried to perform portfolio optimization using the FA. It was observed that the algorithm helped investors make conscious choices between risk and return adapted to certain investment constraints.

There are also studies on the SA (Garcia et al. 2022; Lang et al. 2022; Som and Kayal 2022; Doğan et al. 2024; Lukovac et al. 2017). Garcia et al. (2022) stated that the SA method is effective in portfolio optimization for problems that include constraints such as fixed transaction costs. Lang et al. (2022) tested Simulated, Digital, and Quantum annealing techniques in their study on New York Stock Exchange Exchange Traded Funds. Simulated and digital annealing offer suitable approaches for real-world portfolio optimization. Som and Kayal 2022 performed portfolio optimization with cryptocurrencies, gold, and stocks in ten countries between 2014 and 2020 using the SA. He showed that Bitcoin has an enhancing effect on portfolio performance when the SA method is used. Doğan et al. (2022) researched the SA method on stocks in the BIST30 index. The results obtained show that the algorithm can be used on Borsa Istanbul.

Nutshell, the studies on portfolio optimization of FA and SAs are not at a sufficient level. In addition, no study has been found comparing these two algorithms with the traditional mean–variance model. Studies on Borsa Istanbul are also few. We aim to fill the gap in this field by comparing these algorithms with the classical portfolio optimization solution method and performing a test on Borsa Istanbul.

4 Analysing the Portfolio

The data and methods required for analysis will be explained in this section, All analyses are performed with Python Jupyter Notebook version 7.0.8. The equations for Markowitz's MV model, the FA, and the SA algorithm are given in Appendix.

4.1 Data

The study aims to compare the performances of Markowitz's MV model, the FA, and the SA Model. The data set used for this purpose includes the closing stock prices and annual risk-free interest rates of companies listed in the BIST Dividend 25 Index. The data of the study covering the period between 01/01/2018 and 31/12/2023 were obtained through Data Stream Refinitiv/Eikon. Risk-free interest rates are included in the analysis by taking the Government Domestic Debt Securities (DIBS) data covering the period 2018–2023 on an annual basis.

Annual returns are calculated by taking the daily stock closing prices from the data set. Annual performance data are obtained by normalizing the stock returns for each year. The obtained metrics are compared and the performances of the traditional method (the MV method) and the meta-heuristic artificial intelligence models (the FA and the SA) are compared.

4.2 Performance Metrics

Performance metrics include expected return, portfolio risk (volatility), Sharpe ratio, coefficient of variation (beta), and downside risk. Expected return, which is the basic indicator of finance and the main determinant of risk premium (Martin 2017:367); portfolio risk, which represents the riskiness of all securities for optimum diversification within the portfolio (Bartram and Dufey 2001: 89); Sharpe Ratio, which is the ratio of the expected return and the risk-free interest rate difference (excess expected return) to the standard deviation (Lo 2002:36); beta coefficient, which is calculated by dividing the covariance of the market excess return and the stock excess return by the market variance as an indicator of systematic risk (Fabozzi and Francis 1978:102), and finally downside risk, which is the risk that the realized return will be below the expected return or the uncertainty about the size of this difference (Grootveld and Hallerbach 1999:306) are used as performance indicators in this study.

4.3 Metaheuristic Artificial Intelligence Models

Both FA and SA attempt to achieve a risk-return balance in portfolio optimization. FA is inspired by the light-emitting behavior of social insects (fireflies) and allows it to move between solutions according to light intensity. It is a population-based stochastic search technique, and in the algorithm, each firefly is randomly placed in the search area of the target function (Wang and Song 2024: 7).[1] If the brightness of the fireflies is different, they will behave towards each other; if their brightness is similar, they will behave randomly (Zain et al. 2013:513). Although each of them has a different light level, when viewed holistically, they act in an organized manner and are used in solving non-linear optimization problems (Senthilnath et al. 2011:164).

SA is a generally applicable and easy-to-implement probabilistic and heuristic approximation algorithm that can produce good solutions for an optimization problem (Rutenbar 1989:19; Bertsimas and Tsitsiklis 1993:13). SA is derived by taking inspiration from a thermodynamic process (Crama and Schyns 2003:551). It can avoid local minima while searching for a global minimum. Simulated annealing starts with a randomly selected initial solution and then a neighbor of this solution is generated by a mechanism and the change in cost is calculated. If a decrease in cost is found, the current solution is replaced with the generated neighbor, otherwise the current solution is kept. The process is repeated until no more improvements are found in the vicinity of the current solution, so the local search algorithm ends at a local minimum (Lai and Chan 1997:115).

[1] The Wang and Song (2024) can be viewed for more detailed characteristics of FA.

4.4 Results of the Analyses

The findings of the study are presented in Table 1. In terms of returns, the Firefly and SA portfolios generally have higher return rates than average. Specifically, in 2020 and 2022, the Firefly portfolio achieved significant gains. It can also be said that the mean–variance portfolio generally achieves better returns compared to the SA portfolio. Portfolios that provide high returns can often be risky. The high returns in 2020 and 2022 may be due to price fluctuations caused by COVID-19. High returns may be attractive depending on the risk tolerance of the portfolios.

When assessing the findings in Table 1 in terms of risk, the Firefly portfolio generally exhibits higher volatility, particularly in 2020 and 2023. The SA portfolio also shows high volatility, with even higher risk in 2023. The Mean–Variance portfolio usually has lower volatility, with notably lower risk in 2019 and 2022. Higher volatility can lead to potentially greater gains but also bigger losses. You need to be able to tolerate high volatility depending on your portfolio's risk level.

The Firefly portfolio generally has the highest Sharpe ratio, especially in 2019 and 2020. The SA portfolio also performs well, but its Sharpe ratio declined in 2021. The mean–variance portfolio shows more variable performance, with high Sharpe ratios in some years and low in others. The Sharpe ratio measures return per unit of risk. A high Sharpe ratio indicates good risk management and high returns relative to risk. The Firefly portfolio typically demonstrates the best performance in terms of risk-adjusted returns.

The Firefly portfolio had low beta values in 2018 and 2020, indicating a lower correlation with the market. However, higher beta values were observed in 2021 and 2022. The SA portfolio showed high beta values in 2018 and 2021, signifying a higher correlation with market movements. The mean–variance portfolio generally had variable beta values, showing a high correlation with the market in some years and low in others.

The Firefly portfolio generally has low downside risks, particularly in 2019 and 2022. The SA portfolio shows low downside risk in 2022, but higher risk in other years. The Mean–Variance portfolio typically has high downside risks, especially in 2021 and 2023. Downside risk measures the risks arising from adverse price movements. Lower downside risk indicates protection against loss of investment value. The Firefly portfolio generally performs well in this regard.

5 Conclusion

We presented a contrast to the traditional Markowitz MV model with Firefly and SAs on the BIST Dividend 25 Index. The Firefly portfolio generally has the highest return and Sharpe ratio with high volatility and variable beta values. This portfolio might be suitable for investors with a high-risk tolerance. The SA portfolio also provides high returns but has higher downside risks and beta values. Downside risk represents

Table 1 Findings

Year	Performance metrics	FA	SA	MV
2018	Expected return	− 59.79%	− 36.08%	− 14.24%
	Portfolio risk	25.15%	20.56%	13.91%
	Sharpe ratio	− 2.98	− 1.76	− 2.12
	Coefficient of variation (beta)	− 0.42	0.84	0.51
	Downside risk	1.82%	16.67%	15.36%
2019	Expected return	56.87%	43.11%	53.57%
	Portfolio risk	18.61%	20.31%	13.79%
	Sharpe ratio	2.24	2.11	2.78
	Coefficient of variation (beta)	0.33	0.93	0.49
	Downside risk	1.27%	13.77%	14.17%
2020	Expected return	110.22%	37.90%	61.25%
	Portfolio risk	31.96%	25.48%	20.55%
	Sharpe ratio	3.06	1.48	2.38
	Coefficient of variation (beta)	0.29	0.06	0.04
	Downside risk	2.32%	18.60%	22.56%
2021	Expected return	71.87%	30.37%	17.25%
	Portfolio risk	27.19%	24.44%	22.96%
	Sharpe ratio	2.02	1.24	0.01
	Coefficient of variation (beta)	0.38	0.92	0.83
	Downside risk	2.07%	18.66%	27.37%
2022	Expected return	122.24%	110.21%	108.73%
	Portfolio risk	26.74%	27.47%	24.57%
	Sharpe ratio	3.91	4.01	3.71
	Coefficient of variation (beta)	0.22	0.93	0.78
	Downside risk	1.90%	17.55%	27.84%
2023	Expected return	76.04%	51.68%	41.67%
	Portfolio risk	38.87%	39.76%	32.83%
	Sharpe ratio	1.52	1.30	0.75
	Coefficient of variation (beta)	0.51	0.99	0.81
	Downside risk	2.26%	25.07%	34.20%

the worst-case scenario and findings on SA can guide investors. The Mean–Variance portfolio has lower volatility and downside risk, but its Sharpe ratio and returns are generally lower compared to the other portfolios. The best portfolio depends on the investor's risk tolerance and return expectations. Those seeking high risk and returns might prefer the Firefly portfolio, while those favoring stability and lower risk might choose the MV portfolio. The SA portfolio could be suitable for investors with a moderate risk tolerance. The findings obtained from the study are consistent with the studies indicating that the FA performs better than the traditional MV method (Tuba and Bacanin 2014; Heidari and Neshatizadeh 2018) and the SA performs better than the traditional MV method (Garcia et al. 2022; Lang et al. 2022; Som and Kayal 2022; Doğan et al. 2024; Lukovac et al. 2017).

Investors need investment strategies that will provide them with returns excess market returns to earn more than other investors. Since everyone can access public information, it becomes difficult to achieve excess market return. Here, the importance of diversification emerges. Diversification reduces risk by balancing individual risk factors. The BIST Dividend 25 Index that we used in our study includes many sectors such as banking, automotive, durable consumer goods, construction, and pharmaceuticals. These sectors are included in the portfolios at different weight levels determined by the algorithms. With these calculations, diversification is created with the best performance.

Algorithms may not provide answers to all questions. An algorithm that performs well on certain data sets may perform less well on another data set. Metaheuristic algorithms have generally been used in developed markets and their validity has been proven. We tested them in Borsa Istanbul, a developing market, and their validity has been proven. What should be understood from this is that the perspective on algorithms should be cautious, and they are also promising for future studies. Increasing sample sizes and the number of studies will allow us to obtain more comprehensive information about algorithms. Future research can be done with extensions of FA and Quantum and Digital Annealing algorithms. Also, new structuring can be done by adding skewness and kurtosis values to the MV model. Thus, obstacles caused by the constraints related to portfolio optimization can be overcome.

Appendix

Markowitz Mean Variance Optimization

The MV model max. and min. formulation can be written as (Zhang et al. 2018:126):

$$\begin{cases} \max E[\xi_1 x_1 + \xi_2 x_2 + \ldots + \xi_n x_n] \\ s.t. Var[\xi_1 x_1 + \xi_2 x_2 + \ldots + \xi_n x_n] \leq \beta \\ x_1 + x_2 + \ldots + x_n = 1 \\ x_i \geq 0, i = 1, 2 \ldots, n, \end{cases}$$

$$\begin{cases} \min Var[\xi_1 x_1 + \xi_2 x_2 + \ldots + \xi_n x_n] \\ s.t. E[\xi_1 x_1 + \xi_2 x_2 + \ldots + \xi_n x_n] \geq \alpha \\ x_1 + x_2 + \ldots + x_n = 1 \\ x_i \geq 0, i = 1, 2 \ldots, n, \end{cases}$$

E is the expected value, Var is the variance, x_i is the monetary amount in the i security, ξ i is the random return for the i-th security, and β is the maximum level of risk that an investor can accept. α represents the minimum return on investment that the investor can accept.

Firefly Algorithm

For a D-dimensional search space and a total population size NP in the population, the i-th firefly (i = 1, 2,..., j, NP) is shown as in the equation below (Wang and Song 2024):

$$X_i = (x_{i1}, x_{i2}, \ldots, x_{iD})$$

In nature, fireflies rely on the intensity of their own light to attract potential mates or companions. The attraction (βr) between two fireflies is mainly determined by their light intensities, which can be expressed using in the equation below:

$$\beta(r) = \beta_0 * e^{-y * r_{i,j}2}$$

g is a constant light absorption coefficient, β_0 is the attraction at r = 0, the distance $r_{i,j}$ between any two fireflies i and j at X_i and X_j can be calculated according to the Euclidean distance shown in the equation below:

$$r_{i,j} = X_i - X_j = \sqrt{\sum_{d=1}^{D} \left(x_{i,d} - x_{j,d}\right)^2}$$

For fireflies x_i and x_j, if the brightness of firefly x_j is greater than the brightness of firefly x_i, firefly x_i will move towards firefly x_j. The position update formula is:

$$x_i^{t+1} = x_i^t + \beta_0 e^{-\gamma r_{i,j}^2}\left(x_j^t - x_i^t\right) + \alpha\left(rand - \frac{1}{2}\right)$$

a is called the step factor, rand represents a random number uniformly distributed in the interval [0, 1].

Simulated Anneling Algorithm

SA is a technique for combinatorial optimization problems, such as minimizing functions of many variables (Rutenbar 1989). SA is shown in the equation below (Ghannadi et al. 2023):

$$p = \begin{cases} 1 & if\ E(X_{new}) < E(X_{old}) \\ exp^{(\frac{E(X_{new})-E(X_{old})}{T}} & if\ E(X_{new}) \geq E(X_{old}) \end{cases}$$

T is the temperature, E(Xnew) and E(Xold) represent the energy of the system in the Xnew and Xold states, respectively.

References

Abdel-Basset M, Abdel-Fatah L, Sangaiah AK (2018) Metaheuristic algorithms: a comprehensive review. In: Sangaiah AK, Sheng M, Zhang Z (eds) Computational ıntelligence for multimedia big data on the cloud with engineering applications, pp 185–231. https://doi.org/10.1016/b978-0-12-813314-9.00010-4

Aït-Sahalia Y, Amengual D, Manresa E (2015) Market-based estimation of stochastic volatility models. J Econ 187(2):418–435. https://doi.org/10.1016/j.jeconom.2015.02.028

Avramov D, Zhou G (2010) Bayesian portfolio analysis. Annu Rev Financ Econ 2(1):25–47. https://doi.org/10.1146/annurev-financial-120209-133947

Bailey WJ, Masson R, Raeside R (1998) Choosing successful technology development partners: a best-practice model. Int J Technol Manage 15(1/2):124–124. https://doi.org/10.1504/ijtm.1998.002592

Bartram SM, Dufey G (2001) International portfolio investment: theory, evidence, and institutional framework. Financ Mark Inst Instrum 10:85–155. https://doi.org/10.1111/1468-0416.00043

Baumöl WJ (1966) Mathematical analysis of portfolio selection: principles and application. Financ Anal J 22(5):95–99. https://doi.org/10.2469/faj.v22.n5.95

Behr P, Guettler A, Miebs F (2013) On portfolio optimization: imposing the right constraints. J Bank Finance 37(4):1232–1242. https://doi.org/10.1016/j.jbankfin.2012.11.020

Bertsimas D, Tsitsiklis J (1993) Simulated Annealing. Stat Sci 8(1):10–15. http://www.jstor.org/stable/2246034

Boussaïd I, Lepagnot J, Siarry P (2013) A survey on optimization metaheuristics. Inf Sci 237:82–117. https://doi.org/10.1016/j.ins.2013.02.041

Butler R, Lambson VE (2018) The simplest non-expected utility model for lottery and portfolio choices. Asia-Pac J Risk Insur 12(1). https://doi.org/10.1515/apjri-2017-0006

Cai T, Hu J, Li Y, Zheng X (2020) High-dimensional minimum variance portfolio estimation based on high-frequency data. J Econ 214(2):482–494. https://doi.org/10.1016/j.jeconom.2019.04.039

Chaves DB, Hsu JC, Li F, Shakernia O (2011) Risk parity portfolio vs. other asset allocation heuristic portfolios. J Invest 20(1):108–118. https://doi.org/10.3905/joi.2011.20.1.108

Constantinides GM, Malliaris AG (1995) Portfolio theory. Handb Oper Res Manag Sci 9:1–30. https://doi.org/10.1016/s0927-0507(05)80045-3

Crama Y, Schyns M (2003) SA for complex portfolio selection problems. Eur J Oper Res 150(3):546–571. https://doi.org/10.1016/S0377-2217(02)00784-1

Doğan S, Bezgin MS, Karaçayır E (2024) The portfolio optimization with simulated annealing algorithm: an application of Borsa Istanbul. Gazi Iktisat Ve Işletme Derg 10(1):1–15. https://doi.org/10.30855/gjeb.2024.10.1.001

Draviam T, Chellathurai T (2002) Generalized Markowitz mean–variance principles for multi–period portfolio–selection problems. Proc R Soc Lond A 458:2571–2607. https://doi.org/10.1098/rspa.2002.0983

Fabozzi FJ, Markowitz HM, Kolm PN, Gupta F (2012) Mean-variance model for portfolio selection. In: Fabozzi FJ (ed) Encyclopedia of financial models. https://doi.org/10.1002/9781118182635.efm0003

Fabozzi FJ, Francis JC (1978) Beta as a random coefficient. J Financ Quant Anal 13(1):101–116. https://doi.org/10.2307/2330525

Fernández-Navarro F, Martínez-Nieto L, Carbonero-Ruz M, Montero-Romero T (2021) Mean squared variance portfolio: a mixed-integer linear programming formulation. Mathematics 9(3):223. https://doi.org/10.3390/math9030223

Garcia RÁ, García-Ripoll JJ, Porras D (2022) Portfolio optimization with discrete simulated annealing. arXiv preprint arXiv:2210.00807

Ghannadi P, Kourehli SS, Mirjalili S (2023) A review of the application of the SA in structural health monitoring (1995–2021). Frat Ed Integrità Strutt 17(64):51–76. https://doi.org/10.3221/IGF-ESIS.64.04

Grootveld H, Hallerbach W (1999) Variance vs downside risk: is there really that much difference? Eur J Oper Res 114(2):304–319. https://doi.org/10.1016/S0377-2217(98)00258-6

Heidari H, Neshatizadeh L (2018) Stock portfolio-optimization model by mean-semi-variance approach using of firefly algorithm and imperialist competitive algorithm. Int J Bus Dev Stud 10(1):115–143

Hurley WJ, Brimberg J (2015) A note on the sensitivity of the strategic asset allocation problem. Oper Res Perspect 2:133–136. https://doi.org/10.1016/j.orp.2015.06.003

Jensen MC (1972) Capital markets: theory and evidence. Bell J Econ Manag Sci 3(2):357–398. https://doi.org/10.2307/3003029

Jobson JD, Korkie B (1980) Estimation for markowitz efficient portfolios. J Am Stat Assoc 75(371):544–554. https://doi.org/10.1080/01621459.1980.10477507

Johari NF, Zain AM, Noorfa MH, Udin A (2013) FA for optimization problem. Appl Mech Mater 421:512–517. https://doi.org/10.4028/www.scientific.net/amm.421.512

Kan R, Zhou G (2007) Optimal portfolio choice with parameter uncertainty. J Financ Quant Anal 42(3):621–656. https://doi.org/10.1017/s0022109000004129

Kaplan PD, Siegel LB (1994) Portfolio theory is still alive and well. J Invest 3(3):45–46. https://doi.org/10.3905/joi.3.3.45

Keller WJ, Putten HV (2013) Tactical MPT and momentum: the modern asset allocation (MAA). Available at SSRN. https://doi.org/10.2139/ssrn.2373086

Kolm PN, Tütüncü R, Fabozzi FJ (2013) 60 years of portfolio optimization: practical challenges and current trends. Eur J Oper Res 234(2):356–371. https://doi.org/10.1016/j.ejor.2013.10.060

Lai KK, Chan JWM (1997) Developing a simulated annealing algorithm for the cutting stock problem. Comput Ind Eng 32(1):115–127. https://doi.org/10.1016/S0360-8352(96)00205-7

Lang J, Zielinski S, Feld S (2022) Strategic portfolio optimization using simulated, digital, and quantum annealing. Appl Sci 12(23):12288. https://doi.org/10.3390/app122312288

Lazulfa I (2019) A firefly algorithm for portfolio optimization. J Indones Math Soc 25(3):282–291. https://doi.org/10.22342/jims.25.3.821.282-291

Lo AW (2002) The statistics of sharpe ratios. Financ Anal J 58(4):36–52. https://doi.org/10.2469/faj.v58.n4.2453

Lukovac V, Pamučar D, Popović M, Đorović B (2017) Portfolio model for analyzing human resources: an approach based on neuro-fuzzy modeling and the SA. Expert Syst Appl 90:318–331. https://doi.org/10.1016/j.eswa.2017.08.034

Lyle MR, Yohn TL (2021) Fundamental analysis and mean-variance optimal portfolios. Account Rev 96(6):303–327. https://doi.org/10.2308/tar-2019-0622

Markowitz HM (1952) Portfolio selection. J Financ 7(1):77–91. https://doi.org/10.1111/j.1540-6261.1952.tb01525.x

Markowitz HM (1991) Foundations of portfolio theory. J Financ 46(2):469–477. https://doi.org/10.1111/j.1540-6261.1991.tb02669.x

Markowitz HM, Dijk ELV (2003) Single-period mean-variance analysis in a changing world (corrected). Financ Anal J 59(2):30–44. https://doi.org/10.2469/faj.v59.n2.2512

Markowitz HM (1999) The early history of portfolio theory: 1600–1960. Financ Anal J 55(4):5–16. https://www.jstor.org/stable/4480178

Martin I (2017) What is the expected return on the market? Q J Econ 132 (1):367–433. https://doi.org/10.1093/qje/qjw034

Mazgit İ (2013) Endeks Kapsamında Olmanın Hisse Senedi Getirilerine Etkisi: BIST Temettü 25 Endeksi Üzerine Bir Uygulama. Sosyoekonomi 20(20). https://doi.org/10.17233/se.56199

Monticelli AJ, Romero R, Asada EN (2007) Fundamentals of SA. In: Lee KY, El-Sharkawi MA (eds) Modern heuristic optimization techniques: theory and applications to power systems, pp 123–146. https://doi.org/10.1002/9780470225868.ch7

Nisani D (2018) Portfolio selection using the riskiness index. Stud Econ Financ 35(2):330–339. https://doi.org/10.1108/sef-03-2017-0058

Rubinstein M (2002) Markowitz's "portfolio selection": a fifty-year retrospective. J Financ 57(3):1041–1045. https://doi.org/10.1111/1540-6261.00453

Rutenbar RA (1989) SAs: an overview. IEEE Circuits Devices Mag 5(1):19–26. https://doi.org/10.1109/101.17235

Senthilnath J, Omkar SN, Mani V (2011) Clustering using firefly algorithm: performance study. Swarm Evol Comput 1(3):164–171. https://doi.org/10.1016/j.swevo.2011.06.003

Som A, Kayal P (2022) A multicountry comparison of cryptocurrency vs gold: portfolio optimization through generalized SA. Blockchain: Res Appl 3(3). https://doi.org/10.1016/j.bcra.2022.100075

Stoyanov SV, Rachev ST, Fabozzi FJ (2007) Optimal financial portfolios. Appl Math Financ 14(5):401–436. https://doi.org/10.1080/13504860701255292

Sutiene K, Schwendner P, Sipos C, Lorenzo L, Mirchev M, Lameski P, Cerneviciene J (2024) Enhancing portfolio management using artificial intelligence: literature review. Front Artif Intell 7. https://doi.org/10.3389/frai.2024.1371502

Tuba M, Bacanin N (2014) Upgraded FA for portfolio optimization problem. In: UKSim-AMSS 16th ınternational conference on computer modelling and simulation, Cambridge, UK, pp 11–118. https://doi.org/10.1109/uksim.2014.25

Wang Y, Song S (2024) Detection of sweet corn seed viability based on hyperspectral imaging combined with FA optimized deep learning. Front Plant Sci 15:1361309. https://doi.org/10.3389/fpls.2024.1361309

Xiang Y, Gubian S, Suomela B, Hoeng J (2013) Generalized simulated annealing for global optimization: the GenSA package. R J 5(1):13. https://doi.org/10.32614/RJ-2013-002

Xiong JX, Idzorek TM (2011) The impact of skewness and fat tails on the asset allocation decision. Financ Anal J 67(2):23–35. https://doi.org/10.2469/faj.v67.n2.5

Yang X, He X (2013) FA: recent advances and applications. Int J Swarm Intell 1(1):36–50. https://doi.org/10.1504/ijsi.2013.055801

Zhang Y, Li X, Guo S (2018) Portfolio selection problems with Markowitz's mean–variance framework: a review of literature. Fuzzy Optim Decis Making 17:125–158. https://doi.org/10.1007/s10700-017-9266-z

Zhu H, Wang Y, Wang K, Chen Y (2011) Particle swarm optimization (PSO) for the constrained portfolio optimization problem. Expert Syst Appl 38(8):10161–10169. https://doi.org/10.1016/j.eswa.2011.02.075

9783031832659